T0136321

SURROGATE BROODSTOCK DEVELOPMENT IN AQUACULTURE

The rapidly changing climatic condition coupled with habitat destruction, aquatic pollution and increasing anthropogenic pressure on water bodies have resulted in decline of many important fish population and some of them even become endangered. As of now the breeding protocol for seed production in captivity is developed for only handful of fish species and mostly their seed is collected from natural resources for aquaculture. This factor limits the efforts for species diversification in aquaculture. There are approaches/ technologies to generate seed of such fish species for aquaculture, especially the species that are too large to propagate in captivity or species those do not response to hormonal treatments due to stress of confinement. One of the viable approach is surrogate broodstock development using adult fish as the recipient. The obvious advantage of using adult fish as recipient is that, the donor-derived gametes can be generated within few months after stem cell transplantation; oppose to using embryos or young hatchlings those take years together to attain sexual maturity.

Dr. Sullip Kumar Majhi is working as Principal Scientist in Fish Conservation Division of ICAR-National Bureau of Fish Genetic Resources, Lucknow. He constantly proved excellence in developing assisted reproductive technology in fish for conservation and propagation of elite germplasm. He started his career in the year 2003 in Indian Council of Agricultural Research (ICAR) as scientist and having 16 years of research experience. He has organized many training program and workshops to build skilled manpower in the field of reproductive biotechnology and assisted reproductive technology.

SURROGATE BROODSTOCK DEVELOPMENT IN AQUACULTURE

Sullip Kumar Majhi

ICAR-National Bureau of Fish Genetic Resources,
Canal Ring Road, Telibagh,
Dilkhusa P.O., Lucknow 226 002

CRC Press
Taylor & Francis Group
Boca Raton London New York

CRC Press is an imprint of the
Taylor & Francis Group, an **informa** business

NARENDRA PUBLISHING HOUSE
DELHI (INDIA)

First published 2021
by CRC Press
2 Park Square, Milton Park, Abingdon, Oxon, OX14 4RN

and by CRC Press
6000 Broken Sound Parkway NW, Suite 300, Boca Raton, FL 33487-2742

British Library Cataloguing-in-Publication Data
A catalogue record for this book is available from the British Library

Library of Congress Cataloging-in-Publication Data
A catalog record has been requested

ISBN: 978-0-367-56404-9 (hbk)
ISBN: 978-1-003-09762-4 (ebk)

**NARENDRA PUBLISHING
HOUSE
DELHI (INDIA)**

Contents

Preface

The rapidly changing climatic condition coupled with habitat destruction, aquatic pollution and increasing anthropogenic pressure on water bodies have resulted in decline of many important fish population and some of them even become endangered. As of now the breeding protocol for seed production in captivity is developed for only handful of fish species and mostly their seed is collected from natural resources for aquaculture. This factor limits the efforts for species diversification in aquaculture. There are approaches/ technologies to generate seed of such fish species for aquaculture, especially the species that are too large to propagate in captivity or species those do not response to hormonal treatments due to stress of confinement. One of the viable approach is surrogate broodstock development using adult fish as the recipient. The obvious advantage of using adult fish as recipient is that, the donor-derived gametes can be generated within few months after stem cell transplantation; oppose to using embryos or young hatchlings those take years together to attain sexual maturity. There are certain criteria to chose the recipient fish for surrogate broodstock development. For example, the fish species chosen as recipient should be easy to breed in captivity, closely related to the donor fish, attain maturity in months time and should be a prolific breeder. In this book, detail methodology for surrogate broodstock development are discussed, including strategy to ablate endogenous germ cells in adult fish and method of donor germ cell transplantation into the gonads of recipient fish. It is expected that, the book will be helpful to the aquaculture industry especially the seed producing units. Also, researchers and academicians will be greatly benefited from this book as the contents are explained in simple understandable language.

Dr. Sullip Kumar Majhi

CHAPTER-1
■———■
GENERAL INTRODUCTION

Germ Cell Transplantation was first demonstrated in mouse and was performed with dispersed testicular cell suspensions containing unknown numbers of spermatogonia cells derived from donor males and microinjected into sterilized male recipients, leading to establishment of donor-derived spermatogenesis and production of surrogate gametes. Since then, the technique has been extensively used for the purpose of basic research, reproductive medicine and treatment of infertility, but surprisingly remains, in part, unexplored in fish, even though germ cell transplantation has potential applications in fish bioengineering (see Fig. 1)

During the beginning of 20th century, making a breakthrough, a similar approach was developed by the Japanese scientists in fish using transplantation of primordial germ cells carrying green fluorescent protein into coelomatic cavity of rainbow trout hatchlings resulted in production of sperm with donor genetic characteristics that was capable of fertilizing the eggs. Further, using the same methodology, xenogenic transplantation between rainbow (*Onchorhynchus mykiss*) and masu (*O.masu*) trout were also successfully performed . However, germ cell transplantation using embryos and/or hatchlings requires very sophisticated instruments for the primordial germ cells isolation and quantum of labor for cell transplantation.

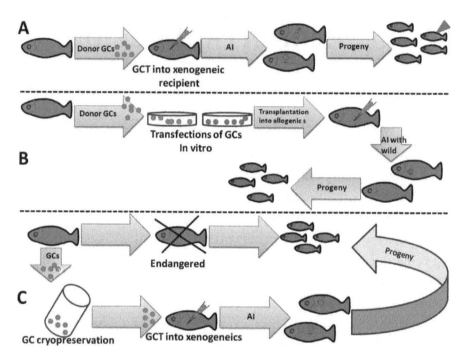

Fig. 1. Potential applications of germ cell transplantation in the field of reproductive biotechnology. A) Production of fish from surrogate parents. Using this technique, commercially important species can be quickly propagated by transplanting germ cells from target species into xenogeneic recipients. B) Transgenic production. This technique can be used by transfecting germ cells *in vitro* and transplanted into allogeneic recipients, followed by retrieval of germ cells. C) Preservation of genetic materials of endangered species. The cryopreservation and xenogeneic transplantation of germ cells from endangered species make it possible to regenerate them, even in case of extinction.

Beside, the transplanted embryos takes considerably long time to reach adulthood and to produce the donor-derived functional gametes, for which significant amount of investment will be required for feeds and manpower inputs. Consequently, the hatchery units, which are the end user of the technique may not be interested to adopt the technique for the commercial production of valued fish seeds. In contrast, development of surrogate broodstock involving transplantation of spermatogonia cells derived from target species into a related adult fish species, depleted of endogenous germ cells using suitable ablative strategy and, for which captive breeding technique are well developed, might considerably shorten in production of donor-derived gametes and, make the technique of germ cell transplantation more simple and viable to be practically feasible for end users.

In this context, in this book, germ cell transplantation in adult fish is systematically narrated using two congeneric model species of atherinopsid fishes, the Pejerrey (*Odontesthes bonariensis*) and the Patagonian Pejerrey (*O. hatcheri*) as donor and recipient, respectively. These two species used as model due to the wealth of basic information available on its reproductive physiology, can be easily bred in captivity and availability of several genetic markers to distinguish them.

This book is composed of five chapters as follows: 1) Recipient preparation for surrogate broodstock development, 2) Surrogate broodstock development by surgical approach, 3) Surrogate broodstock development by non-surgical approach, 4) Importance of surrogate broodstock in aquaculture and 5) Instrument used for surrogate broodstock development in fish. These chapters aim to explore in depth from ablation strategy of endogenous germ cells in recipients prior to introduction of donor germ cells for production of donor-derived progeny from the surrogate parents by artificial fertilization and natural spawning.

It is anticipated that the contents of this book will have broader implications in aquaculture for production of progeny from the fish species that are difficult to propagate in captivity but posses high commercial value, which are too large for hatchery rearing and that do not spawn due to the stress of confinement, or whose maturation cycle is associated with complex migratory behavior.

CHAPTER-2

■———————■

RECIPIENT PREPARATION FOR SURROGATE BROODSTOCK DEVELOPMENT

G erm cell transplantation, a technique first demonstrated in mouse, is a promising reproductive technology with potential application in animal reproduction and conservation of endangered species from fish to mammals. This procedure, which is based on the transplantation of donor germ cells into a surrogate gonad, allows the speedy and theoretically unlimited production of donor-origin gametes from the donor without interference from the host's genome, that is, the animals produced with transplanted cells are not transgenic animals. Thus, highly productive individuals could be propagated far above their reproductive capacity and beyond their lifespan without resorting to time-consuming procedures of gamete collection, cryopreservation, and artificial fertilization. In addition, germ cells could be harvested from immature as well as aging donor animals and can be induced to develop into functional gametes timely by transplantation into an adult, sexually-competent recipient animal. In this way, this procedure offers a way to preserve genetic material from animals that might otherwise die prior to sexual maturity or become senile, and this is extremely important in the case of rare, endangered and commercially important species. There has been recently a great interest in the application of germ cell transplantation to aquatic animals, in particular for fish. Fishes are generally prolific and germ cell transplantation using related species as recipients could be used to propagate commercially important species which are too large for hatchery rearing, that do not spawn due to the stress of confinement, or whose maturation cycle is associated with complex migratory behavior which cannot be reproduced in captivity.

One of the preconditions for surrogate broodstock development is the availability of a suitable host species whose gonads are physiologically and immunologically compatible with the donor's germ cells and which meets the requirements of easy maintenance and breeding in captivity. Closely related species generally meet the physiological and immunological requirements, but the recipient animal preferably should be artificially depleted of its endogenous germ cells so that there is a relatively higher yield of donor-derived gametes. For instance, Brinster and colleagues showed that the introduced cells have improved access to the basal compartment of seminiferous tubules of mouse when the recipient testis is depleted of endogenous germ cells prior to transplantation. Several options for depletion of germ cells have been tested in mammals such as treatment with cytotoxic drugs e.g. Busulfan, irradiation, cold ischemia and hyperthermic treatment. Some of these methods have been also tried on fish and conditions for rapid and efficient depletion of germ cells with general applicability have been established. An alternative method is to perform transplantation into triploid gonads as triploid fish are generally, though not always, considered as sterile. While ingenious, however, this strategy requires the long-term rearing of recipient fishes until the adult size as triploids can be produced only by chromosome manipulation performed shortly after fertilization or by crossing between diploid and tetraploid fish. On the contrary, thermo-chemical treatments protocols for sterilization of fish gonads allow application in cases that require immediate attention such as species facing imminent extinction and for which suitable triploid hosts are not immediately available.

In this context, suitability of treatments with Busulfan and warm water to induce the depletion of endogenous germ cells in adult Patagonian pejerrey, *Odontesthes hatcheri* was evaluated by Majhi and colleagues. They conducted two separate experiments, the first only with males and the second with males and females, wherein the effectiveness of the treatments was evaluated by gonadosomatic index (GSI) determination, qualitative and quantitative (changes in germ cell number) histological observations, and the ability of treated gonads to support gametogenesis. In addition, suitability of the germ cell-specific *vasa* gene transcript level as a molecular tool to monitor the degree of germ cell depletion in treated fish was also examined.

Methods for germ cell depletion

One year old Patagonian pejerrey *Odontesthes hatcheri* (Fig. 1) were used in the experiments (first experiment; only males; mean body weight ± SD of 28.2±10.8 g and second experiment; both sexes; mean body weight ± SD of 23.3 ± 9.3 g for males and 20.5 ± 11.5 g for females). Fish were stocked in 200 L tanks at a density of 6.3 Kg of fish per m^3 and reared in flowing brackish water (0.2-0.5% NaCl) under a constant light cycle (15L9D). Animals were fed with pelleted commercial diet four times per day to satiation. All groups were acclimated for two weeks at 17°C prior to the treatments.

Fig. 1. The model fish species used in the study: *Odontesthes hatcheri*. Scale bar indicates 4 cm.

The first experiment, only with males, tested four combinations of Busulfan (20 and 40 mg/Kg body weight) and temperature (20 and 25°C) whereas a fifth group receiving only the vehicle (DMSO) at 20°C served as a control (hence, groups were named B20T20, B20T25, B40T20, B40T25, and B0T20). In the second experiment, fishes of both sexes were reared at 25°C and received either treatments of 40 mg/Kg Busulfan (B40T25) or only the vehicle (B0T25). They conducted this experiment to confirm the results of the first one, which showed marked germ cell depletion at the highest Busulfan dosage and temperature, to discriminate between the effects of high temperature and the drug, and to ascertain the effects in females. In both experiments, each group was split in two subgroups at four weeks with half of the fish receiving a booster treatment of the same dosage as the previous one. All subgroups were then reared separately until the termination of the experiment at 8 weeks. Fish in the second experiment were reared at 17°C for additional 8 weeks after termination (recovery period) for observation of the permanency/improvement of the histological conditions of the gonads after treatment. Improvement with proliferation of germ cells and gamete formation was taken as an indication of the capacity of the gonads to support gametogenesis. Busulfan dosages were based on published information on effective dosages for mouse. The choice of the temperatures was based on previous findings of Strüssmann *et al.*, 1998, which showed the presence of germ cell deficient fish among *O. hatcheri* exposed to 25°C during the juvenile stage. The cytotoxic drug Busulfan was first dissolved in dimethyl sulfoxide (DMSO) and then further diluted with freshwater fish Ringer solution to make the working concentrations. The limit of solubility for Busulfan in aqueous media is near 4 mg/ml. Thus, to avoid precipitation, it was mixed with fish Ringer solution just before use and maintained at room temperature. For treatment, the fish were anesthetized using 100 ppm of 2-Phenoxyethanol and the body weight and length were recorded. Busulfan was intraperitoneally injected to the fish using a 500 µl micro syringe.

For histological observation, 4-5 fish per sex were randomly sampled from each group at 2, 4 and 8 weeks. Fishes from the second experiment that were reared at 17°C after 8 weeks were sampled also at 12 and 16 weeks. Fish were killed by an overdose (200 ppm) of anesthesia and their body weight was recorded. The gonads were excised, macroscopically examined, photographed using a digital camera, and weighed to the nearest 0.01 g. The middle portion of the right and left gonads from each fish were then immersed in Bouin's fixative for 24 hours and preserved in 70% ethanol. Gonads were processed for light microscopical examination following routine histological procedures up to sectioning at a thickness of 5 μm and staining with hematoxylin-eosin (Annexure-I). About 40-50 serial histological sections from each fish were examined under a microscope at magnifications between 10-60X. The degree of histological degeneration and germ cell loss of each specimen was classified following the set criteria (see Table 1). Digital images taken from five representative histological sections of the right and left lobes of the gonad of each individual in the 2nd experiment were used for determination of the number of spermatogonia or oogonia per section and the cross-section area of the gonad using the Image-Pro Plus software ver. 4.0.

Table 1. Histological criteria for classification of gonadal integrity/degeneration

Males	Class	Females
Conspicuous cysts of spermatogonia and other spermatogenic stages	I	Presence of cysts of oogonia interspersed with oocytes at various stages of development
Only cysts of spermatogonia; efferent ducts may or not contain residual spermatozoa	II	Light hypertrophy of the ovigerous lamellae; oocytes fewer and atretic
Cysts of spermatogonia are few and small	III	Few if any oocytes and greatly reduced number of oogonia
Absence of spermatogonia	IV	Absence of oogonia

The usefulness of *vasa* expression analysis as a tool to monitor the degree of germ cell loss in *O. hatcheri* was examined by *in situ hybridization* (ISH) and Real Time RT-PCR analysis using males from the first experiment. For *vasa* ISH, a piece of testis from the control group was fixed in 4% Paraformaldehyde in PBS (pH 7.4) overnight at 4°C, dehydrated in an ethanol series, and embedded in paraffin. Transversal sections were cut in a microtome to a thickness of 6 μm and mounted onto glass slides. Sections were pre-treated in 1 μg/ml Proteinase K for 10 min (25°C) and then hybridized with 150 μl of digoxigenin-labeled sense and antisense probes at 50°C overnight. *Vasa* probes were synthesized using primers constructed based on a sequence for *O. hatcheri* available in the GenBank (accession number #DQ441593). T7 and T3 phage promoter sequences were appended to the forward (5-GTA ATA CGA CTC ACT ATA GGG CCT CCA ACC AGG GAG CTC ATC AAC C-3) and reverse primers (5-GCA ATT AAC CCT CAC TAA AGG GGT GGC GCT GAA CAT CAG GGT CTG-3), respectively, for direct *in vitro* transcription using T7 and T3 RNA polymerase enzymes. The hybridization signal was detected using Anti-DIG-AP and NBT/BCIP.

Samples for Real Time RT-PCR of *vasa* expression were taken at 8th week from the anterior part of the testes of fishes that received a booster dose on the 4th week and stored in RNA*later* at -80°C until further processing. RNA was extracted using Trizol according to manufacturer's protocol. cDNA was synthesized using oligodT primers and superscript reverse transcriptase. Primers for Real-Time RT-PCR (5-CCT GGA AGC CAG GAA GTT TTC-3 and 5-GGT GCT GAC CCC ACC ATA GA-3) were designed using Primer Express (ver. 2.0). The Real Time PCRs were run in an ABI PRISM

7300 using *Power* SYBR® Green PCR Master Mix in a total volume of 15 µl which included 25 ng of first strand cDNA and 5 pmol of each primer. β-actin was analyzed as an endogenous control. Quantification was performed using the standard curve method with 4 points and the ABI Prism 7300 sequence detection software (v.1.2).

Measured parameters were compared among the treatments by one-way analysis of variance (ANOVA) with Tukey's multiple comparison tests by using Graphpad prism ver.4.00 for windows. Data are presented as mean ± SD and differences between groups were considered as statistically significant at $p<0.05$.

Majhi and colleagues observed that males tolerated well the Busulfan and high temperature treatments, with a minimum survival rate of 88% in the B40T25 group of experiment 1. However, the same treatment caused mortality rates of 20% and 10% in the females at the 6th and 7th weeks, respectively, of experiment 2. No mortality was observed after the 7th week in fishes from either sex in both experiments. The mortality of females was likely due to the toxic effects of Busulfan, as the fish died within 2-3 days of treatment and subsequent appearance of skin ulcerations (Fig. 2). There were no significant changes in gonad-free body weight in any of the treatments compared to the control.

Fig. 2. Side effects of cytotoxic drug Busulfan on fish health. Note appearance of skin ulcerations in the body surface of females (arrows).

The GSI of males from all groups, including the controls, decreased steadily in both experiments between 1 and 8 weeks (Figs. 3 and 4). However, the decreases were much more prominent in the groups treated with the highest Busulfan dosage, especially in the groups receiving a booster treatment at 4 weeks. The testes of males from the B40T25 groups in the 1st and 2nd experiments were visibly shrunken and had a dark coloration at 8th week compared to the testes of control animals (Fig. 5). The microscopic examination revealed active spermatogenesis in all but one male not treated with Busulfan throughout the experiment (Tables 2 and 3; Fig. 6). The exceptional individual, sampled after 8 weeks at 25°C (2nd experiment), had only cysts of spermatogonia and lacked all other spermatogenic stages. Males treated with a single dose of Busulfan lacked cysts of spermatocytes and spermatids but still retained cysts of spermatogonia and some residual spermatozoa in the lumen of the seminiferous tubules and efferent ducts after 2-8 weeks regardless of temperature. Some of these fish also presented conspicuous cysts of abnormal cells with variable size and staining properties ranging from dense basophilic (hematoxylin-stained) to acidophilic (eosin-stained) bodies (Fig. 6). Males treated with a booster dosage of Busulfan at 4 weeks, on the hand, had further degrees of germinal degeneration at 8 weeks and these were by far more severe at the highest temperature (25°C). Determination of the number of spermatogonia per unit area of gonadal cross section in the 2nd experiment corroborated the histological observations, showing progressive and significant losses of germ cells in B40T25 males compared to those from B0T25 (Fig. 7). The testes of males allowed to recover at 17°C after the Busulfan and high temperature treatments showed rapid histological

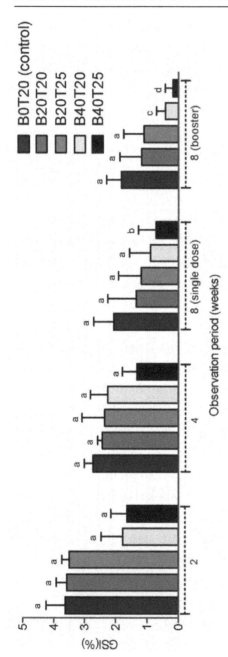

Fig. 3. Changes in the gonado-somatic index of fishes (only males) during experiment 1. Columns with different letters vary significantly (Tukey's multiple comparison test, $p < 0.05$).

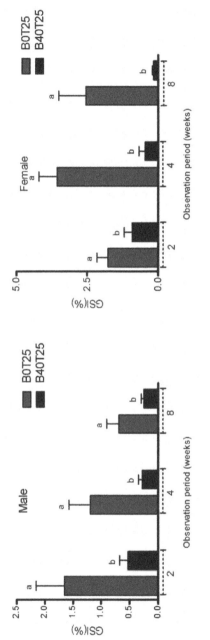

Fig. 4. Changes in the gonado-somatic index of males (left panel) and females (right panel) during experiment 2. Columns with different letters vary significantly (Tukey's multiple comparison test, $p<0.05$).

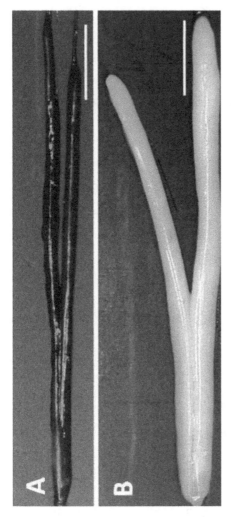

Fig. 5. Macroscopic appearance of testes from Busulfan-high temperature (B40T25; A) and control (B0T25; B) groups after 8 weeks of treatment. Scale bars- 1cm.

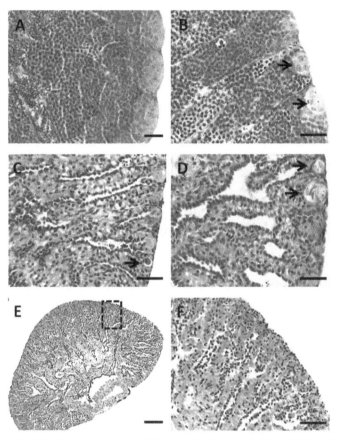

Fig. 6. Histological appearance of the testes in control and Busulfan-high temperature treated groups. A) Testis from a B0T20 male at 0 week showing a thick germinal epithelium and accumulation of spermatozoa in the lumen of the efferent ducts, which indicate active spermatogenesis. B) Testis from a B0T25 male at 8 weeks showing conspicuous cysts of spermatogonia in the periphery of the gonad (arrows), spermatocytes, and spermatids. C) Testis from a B40T25 male at 2 weeks showing a shrunk germinal epithelium with small cysts of spermatogonia and abnormal spermatogonia (arrow). D) Testis from a B40T25 male at 4 weeks showing the greatly atrophied germinal epithelium with only a few remaining spermatogonia (arrows). E) Testis from a B40T25 at 8 weeks that received a booster treatment at 4 weeks showing virtual lack of spermatogonia. F) Detail of the box shown in E. Scale bar-100 μm (A,E), 30 μm (B,C,D,F).

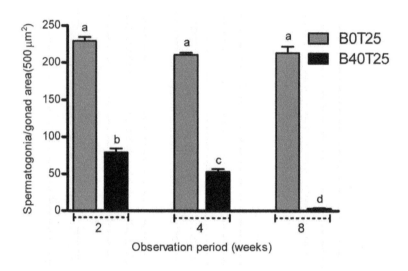

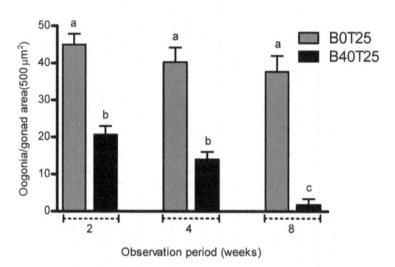

Fig. 7. Changes in the number of spermatogonia and oogonia per unit area of gonadal cross section in control (B0T25) and Busulfan-high temperature (B40T25) groups. Columns with different letters vary significantly (Tukey's multiple comparison test, $p<0.05$).

improvement and re-colonization of germ cells; several individuals even possessed spermatozoa in the lumen of the spermatogenic tubules at 16 weeks (Table 2).

Table 2. Frequency of individuals per category of histological appearance of the testes in adult Patagonian pejerrey (experiment 1)

Treatments	Observation (weeks) period	Number of males			
		Histological category			
		I	II	III	IV
B0T20	2	4	–	–	–
	4	4	–	–	–
	8	2	–	–	–
	8*	2	–	–	–
B20T20	2	–	4	–	–
	4	–	4	–	–
	8	–	2	–	–
	8*	–	2	–	–
B20T25	2	–	4	–	–
	4	–	4	–	–
	8	–	2	–	–
	8*	–	1	1	–
B40T20	2	–	4	–	–
	4	–	4	–	–
	8	–	2	–	–
	8*	–	–	2	–
B40T25	2	–	4	–	–
	4	–	3	1	–
	8	–	2	–	–
	8*	–	–	–	2

* Treated with booster dose

The GSI of females reared at 25°C and not treated with Busulfan in the second experiment showed large individual variation between 2 and 8 weeks whereas that of females treated with the drug showed more consistent and significant decreases (Fig. 8). Females in the B0T25 group did not show any noticeable histological changes throughout the study with the exception of two individuals, one each on the 4th and 8th weeks, whose ovaries showed light atrophy of the ovigerous lamellae, fewer oocytes, and presence of degenerating oogonia (characterized by dense eosin staining; Fig. 8, Table 3). Similar histological characteristics were observed also in B40T25 females on the 2nd and 4th weeks. Moreover, two out of five ovaries sampled from this group on week 4 and three out of five sampled on week 8 had markedly few oogonia and depositions of yellowish-brown pigments, indicating the occurrence of phagocytosis. The remaining two females sampled at 8th week from B40T25 were completely devoid of oogonia in all sections examined (Table 3, Fig. 8). The number of oogonia per unit area of gonad cross section showed a steady decrease in the experimental groups between 2 and 8 weeks (Fig. 7). Females allowed to recover at 17°C showed marked improvement of the histological appearance of the ovaries, presenting few but conspicuous cysts of oogonia in all five individuals examined at 16 weeks and even a few cortical alveoli oocytes in one.

The spatial expression of the *vasa* gene was analyzed by ISH to clarify its localization within the adult testis of Patagonian pejerrey. *Vasa* RNA transcripts were detected mainly in the cytoplasm of spermatogonia, which occupy cysts in the periphery of the gonads (Fig. 9). The signals were very faint or absent in spermatocytes, spermatids, and spermatozoa. The quantitative analysis of *vasa* expression by Real-Time RT-PCR showed that

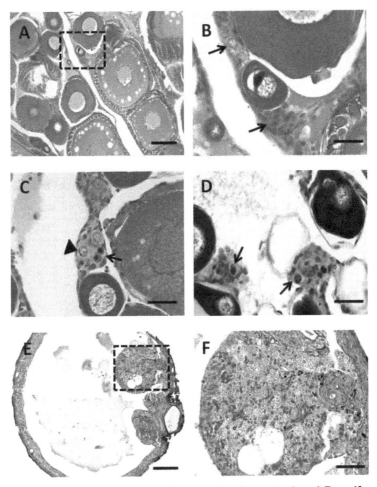

Fig. 8. Histological appearance of the ovaries in control and Busulfan-high temperature treated groups. A) Ovary of a control female at 0 week showing oocytes at various stages of development. B) Detail of the box in 'A' showing prominent cysts of oogonia. C) Ovary from a B0T25 female at 8 weeks showing the presence of normal as well as abnormal (eosinophilic) oogonia (arrow). D) Ovary from a B40T25 female at 2 weeks (note the abundance of abnormal oogonia (arrows). E) Ovary from a B40T25 female at 8 weeks that received a booster treatment at 4th weeks showing the virtual lack of oogonia. F) Detail of the box shown in E. Scale bar-100 μm (A,E), 30 μm (B,C,D,F).

vasa transcript levels were significantly different among treatments ($p<0.05$; Fig. 10). The lowest level of *vasa* transcript was recorded in B40T25 (7.66 ± 1.25), which was 6-fold lower than control males (B0T20; 49.21 ± 6.54). Groups B40T20 (25.24 ± 2.07) and B20T25 (42.12 ± 4.20) had *vasa* expression levels that were intermediate between those at B40T25 and the control.

Table 3. Frequency of individuals per category of histological appearance of the testes and ovaries in adult Patagonian pejerrey (experiment 2)

Treatments	Observation period (weeks)	Sex	Number of fish Histological category			
			I	II	III	IV
B0T25	2	♂	5	–	–	–
		♀	5	–	–	–
	4	♂	5	–	–	–
		♀	4	1	–	–
	8*	♂	4	1	–	–
		♀	4	1	–	–
B40T25	2	♂	–	5	–	–
		♀	–	5	–	–
	4	B&	–	4	1	–
		♀	–	3	2	–
	8*	♂	–	–	–	5
		♀	–	–	3	2
	12RV	♂	3	2	–	–
		♀	2	3	–	–
	16RV	♂	5	–	–	–
		♀	4	1	–	–

* Treated with booster dose; RV Recovery observation

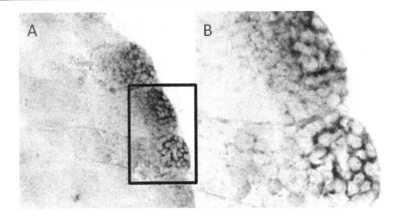

Fig. 9. Expression of *vasa* in adult *O. hatcheri* testis. Antisense RNA transcripts can be observed inside the cysts of spermatogonia (B: detail of the box shown in A). Scale bar-100μm(A), 20μm(B).

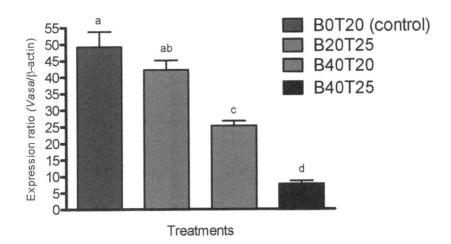

Fig. 10. *Vasa* gene transcript levels at 8 weeks in the testes of adult Patagonian pejerrey that were subjected to different heat and Busulfan treatments. All animals were treated with a booster dosage at 4 weeks. Columns with different letters vary significantly (Tukey's multiple comparison test, $p<0.05$).

Germ cell transplantation technique is considered as a powerful assisted reproductive technology for the conservation of rare or endangered species and also for propagation of commercially important ones that are difficult to breed in hatcheries. Successful germ cell transplantation with production of viable offspring has been performed in fish using normal diploid and triploid salmonid hatchlings as hosts. In these studies, primordial germ cells from donors were transplanted into the body cavity of the hosts to make use of the receptivity of the gonadal primordia to the migrating primordial germ cells during gonadal sex differentiation. Nevertheless, the approach to transplant primary oogonia or spermatogonia from adult donors directly into the gonads of adult hosts by surgical intervention, as demonstrated in mammals, could be an alternative to the technique of using hatchlings when time for obtention of gametes and offspring or the cost of producing suitable hosts is a constraint. Thus, the ability to produce suitable hosts timely and to be able to harvest viable gametes in a relatively short time after germ cell transplantation will add power to this approach and make it even more suitable to respond to sudden threats to endemic fish species such as during environmental cataclysm.

In mammals, chemotherapy, X-ray or Gamma-ray local radiation have been widely used for the purpose of germ cell depletion. In fish, the use of radiation is practically difficult by the fact that the gonads are located in the coelomic cavity and, without surgical approach, it would be almost impossible to effectively irradiate them. Another widely used strategy for depletion of endogenous germ cells is by pretreatment of recipient animals with Busulfan, a cytotoxic drug. However, trials with

Busufan alone have not been as successful to eliminate germ cells in fish as they were in mammals, a fact that could be related to the differences in body temperature between these two taxa. In this context, effectiveness of the chemical treatment with Busulfan in combination with elevated water temperature, an environmental factor that, in itself, causes germ cell degeneration in fish was tested. As described below, the results of this study support the assumption that a combination of both thermo-chemical treatments is more efficient than either thermal or chemical treatment alone.

High temperature (25°C) alone caused a mild regression of the testes and ovaries of *O. hatcheri* after 8 weeks. The observed changes included the disappearance of all intermediate stages of spermatogenesis between spermatogonia and mature sperm in males and shrinkage of the ovary, thickening of the ovarian tunica and incipient hypertrophy of the ovigerous lamella in females. Ito and colleagues also recorded a similar degree of gonadal degeneration in the congeneric species *O. bonariensis* after exposure to 29°C for 4 weeks and observed that this level of regression was equivalent to that observed in wild animals during summer. They concluded that such degeneration represents a natural response to elevated temperature during the annual thermal and reproductive cycle. Nevertheless, further exposure to the same temperature resulted in progressive gonadal degeneration and ultimately sterility. In this context, the histological degeneration seen in *O. hatcheri* after 8 weeks at 25°C presumably represents also an initial response to "summer condition" and yet, continued exposure to this condition might lead to further degeneration. This possibility can be explored because of the obvious advantages of non-chemical methods as well as possible limitations of the chemical treatment with Busulfan.

In contrast to the treatment with temperature alone, Busulfan caused marked gonadal degeneration after 2 weeks. In addition to the observed changes noted in the elevated temperature groups, Busulfan-treated females showed further loss of germ cells in both sexes and intense phagocytic activity by the follicle cells, leading to atrophy of the ovigerous lamellae in females. Histological degeneration of the ovaries, but not that of the testes, involved widespread occurrence of macrophage aggregates, a characteristic reported in ovaries of pejerrey *O. bonariensis* females during prolonged heat stress and other species of fish after exposure to chemical contaminants. Further degeneration in Busulfan-treated groups was clearly dose-dependent, comprising both the nominal dosage and the frequency of injections, and temperature-dependent. Thus, as assumed, the highest dosage and the repeated (booster) treatments, particularly in combination with the highest temperature, resulted in more severe loss of germ cells to the point that some individuals examined at 8 weeks had no noticeable germ cells in any of the histological sections that were analyzed. High temperature and repeated injections were also more effective to deplete germ cells in Nile tilapia, but in that case the investigators attributed the effectiveness of high temperatures to the acceleration of the cell cycle.

Judging by the frequency of animals with few or no germ cells during histological observation and cell counting, and the results of *vasa* expression, male Patagonian pejerrey were more easily depleted of germ cells than females. However, based on the recovery study of the treated fishes at low temperature, no individuals were found to be completely sterile at the end of the treatments. This indicates that, for efficient eradication of germ cells in *O. hatcheri*, the Busulfan treatment period should be extended or combined with higher temperatures. Another alternative could be to increase the dosage or the frequency of treatments with Busulfan. However, it has been reported that high dosages of Busulfan can induce severe bone marrow depression and cause death in mammals. In the case of *O. hatcheri*, it was observed that some individuals developed ulcerations shortly after Busulfan treatment and that mortalities were observed mainly in the groups receiving the highest dosage of the drug. These findings suggest that this dosage (40mg/Kg) is near the tolerance limit of this species. It is noteworthy that Busulfan-induced mortality was more frequent in females than males at the highest temperature. In a previous study, selective mortality of *O. bonariensis* females was also observed during heat-stress compared to males. Thus, in practice, it may be more difficult to completely sterilize females than males of *O. hatcheri*.

The histological observation after a period of recovery at low temperatures revealed that within 8 weeks after the treatment, the gonads were repopulated with germ cells and, in most cases, contained functional gametes. This is an indication that the gonial cells that were not destroyed by the treatment retain the capacity to undergo gametogenesis. It also suggests that somatic cells (such as Sertoli and Leydig cells in males and follicular cell in females, which support the proliferation and development of the germ cells, were not critically affected by the Busulfan and heat treatments. Hence, it was surmised that they would be able to support also transplanted germ cells if these cells are compatible with the host's gonadal environment. Moreover, Busulfan treatment also did not eradicate all endogenous spermatogonia in mammals, leaving some stem cells that would re-initiate spermatogenesis in the host together with the transplanted spermatogonia. This would represent a practical problem for collection of donor-derived gametes as they will have to be separated from those of the host. Still, previous work on mice has shown that the fertility of recipient animals after germ cell transplantation was lower in animals sterilized almost completely by treatment with Busulfan as adults than in recipients prepared by fetal exposure to the drug and where ablation of endogenous germ cell was incomplete.

In summary, the results of Majhi and colleague indicates that a combination of treatments with Busulfan and high temperature might be an effective method to deplete endogenous spermatogonia and oogonia of *O. hatcheri* hosts, and possible in other species, in preparation for surrogate broodstock development. Further improvements could lead to drastic reduction of the endogenous germ cell population, if not complete sterilization of hosts within a span of a few weeks, raising the prospects that such technique might be useful in situations that require immediate action such as during environmental disasters and impending loss of unique genetic materials. Their experiment also demonstrated that, level of the germ cell-specific *vasa* expression can be used to understand the real time insight on size of the remaining germ cell population and can be instrumental for the optimization of protocols for efficient host preparation.

CHAPTER-3

■———■

SURROGATE BROODSTOCK DEVELOPMENT BY SURGICAL APPROACH

Establishment of donor-origin gametogenesis in surrogate animal by germ cell transplantation was pioneered by Brinster and colleagues in 1994. Since then this powerful assisted reproductive technique is widely used for generation of surrogate gametes and progeny. Broadly, it consists on the transplantation of donor germ cells into the gonads of a sterile surrogate animal for rapid and theoretically unlimited production of donor-origin gametes. Over the years, worldwide this technique has gained new scientific interest due to the enormous potential for application in reproductive medicine, preservation of valuable and endangered genetic resources, and animal reproduction. Furthermore, germ cell transplantation has also implications for understanding the regulation of germ cell development and stem cell research.

In the year 2003, a novel approach for germ cell transplantation was developed for use in fish. It relies on the transplantation of primordial germ cells carrying green fluorescent protein into the coelomic cavity of fish hatchlings and these cells were observed to successfully migrate towards and find their way into the germinal ridges. Some years later, Okutsu and colleagues demonstrated that transplantation of germ cells in fish, as originally devised in mammals, does not need to be performed with primordial germ cells as transplantation of spermatogonia cell into hosts also originated donor-derived gametes. These results confirm the technical feasibility and the great potential of this technique in propagation of elite germlines. However, germ cell transplantation using fish embryos and/or hatchlings as recipients requires sophisticated instruments and skills for precise transplantation of donor cells into the body cavity of the small, sometimes only a few millimeters long larvae. In addition, the animals transplanted at these early stages can take a considerably long time to reach adulthood and to produce the donor-derived functional gametes. This adds considerably to the cost of production of offspring from surrogate parents and may be a hindrance to commercial application in hatcheries. In contrast, development of surrogate broodstock involving transplantation of gonial cells (spermatogonia or oogonia) into sexually mature hosts which have been naturally or experimentally depleted of endogenous germ cells might result in almost immediate production of donor-derived gametes. This possibility has been experimentally demonstrated in mammals and was successfully demonstrated by Majhi's group in fish.

They performed intra-gonadal surgical germ cell transplantation between two congeneric model species of atherinopsid fishes as donor and recipient (Fig. 1). The intra-gonadal surgical transplantation of spermatogonia from pejerrey *Odontesthes bonariensis* into Patagonian pejerrey *O. hatcheri*, that were depleted of endogenous germ cells by Busulfan and high water temperature treatments, resulted in re-colonization of recipient testes. Further, the transplanted cells successfully developed into functional sperm and were used in artificial fertilization with eggs of the donor species, generating viable progeny of donor origin.

Methods for Surrogate Broodstock Development

These two species are originally from South America and have been introduced to Southeast Asian countries as candidates for aquaculture. They can be easily bred in captivity and produce viable hybrid offspring, which can be readily distinguished using genetic markers. For recipient preparation, one year old adult; mean body weight ± SD of 37.3 ± 16.8 g male Patagonian pejerrey (*Odontesthes hatcheri*) were first stocked in 200 L tanks at a density of 7.5 kg of fish per m^3 and reared in flowing brackish water (0.2-0.5% NaCl) under a constant light cycle (15L9D). The fishes were acclimated for two weeks at 17°C prior to the treatments for depletion of endogenous germ cells. The pejerrey (*Odontesthes bonariensis*) used as donors of germ cells were 4-5 months old juveniles. Both groups of animals were fed pelleted commercial diet four times per day to satiation.

Fig. 1. The model fish species: *Odontesthes bonariensis* (donor; top) and *O. hatcheri* (recipient; bottom). Scale bar indicates 4 cm.

The endogenous germ cells of Patagonian pejerrey used as recipients were depleted by rearing at a relatively high temperature (25°C) and injected intraperitoneally twice with 40 mg/kg body weight Busulfan at a 4 week interval. Histological observation of gonads confirmed that treated animals were severely depleted of endogenous germ cells 4 weeks after the second injection of Busulfan (Fig. 2). Recipients were used in germ cell transplantation within 7 days after termination of the germ cell depletion treatment.

Donor males were sacrificed by anesthetic overdose and the testes were excised, washed in phosphate-buffered saline (PBS; pH=8.2). The testicular tissue was finely minced and incubated in a dissociating solution containing 0.5% Trypsin (pH 8.2), 5% Fetal Bovine Serum, and 1mM Ca^{2+} in PBS (pH 8.2) for 2 hr at 22°C. The dispersed testicular cells were sieved through a 50μm mesh size nylon screen to eliminate the non-dissociated

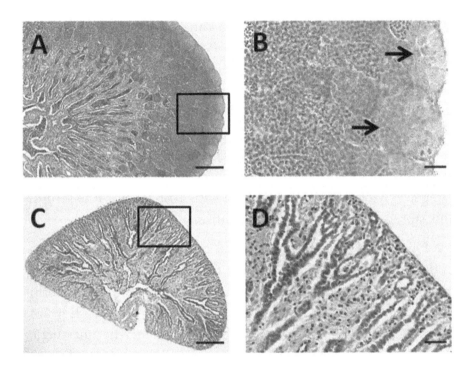

Fig. 2. Histological appearance of the testes of Patagonian pejerrey in control and Busulfan-high temperature treated groups. A,B) Normal testis showing the thick germinal epithelium, the radially-oriented seminiferous lobules, and large cysts of spermatogonia (arrows) in the blind end of the seminiferous lobule (B is a high magnification of the box shown in A). C,D) Testis from the high temperature (25°C) - Busulfan (two injections of 40 mg/Kg 4 weeks apart) treatment group at 8 weeks showing virtual lack of spermatogonia (D is a high magnification of the box shown in C). Scale bars indicates 100 μm (A,C) and 20 μm (B,D).

cell clumps, suspended in discontinuous Percoll gradients of 50%, 25% and 12% and centrifuged at $200 \times g$ for 20 min at 20°C. The bottom phase containing predominantly spermatogonia was selected, the cells were washed two times, and subjected to cell viability test by trypan blue (0.4% w/v) exclusion assay. The cells were then exposed to the CFDA-SE Cell Linker at a concentration of 2 µM (room temperature, 10 minutes) to label the cells for tracking their behavior inside the recipient gonads. The staining procedure was stopped by addition of an equal volume of heat-inactivated fetal bovine serum. Labeled cells were washed three times to remove unincorporated dye, suspended in Dulbecco Modified Eagle Medium with 10% fetal bovine serum, and stored on ice until transplantation.

Thirty one recipients were used for surgical transplantation of donor cell. For this purpose, the fish were anesthetized in 100 ppm 2-phenoxyethanol and placed on an operation platform under a microscope where they received a constant flow of oxygenated, cool water containing 100 ppm of the anesthetic through the gills (Fig. 3). To prevent desiccation, the surface of the fish was moisturized during the entire procedure of cell transplantation, which took about 20 min. per fish on average. An approximately 1.5 cm-long midline incision was made in abdomen and the gonads were carefully lifted from the coelomic cavity and held in place with the help of a sterile soft plastic spatula. A micro syringe and fine glass needle were used to inject the cell suspension carefully into the lumen of each testicular lobe. Each lobe of the testis was injected with 50 µl of cell suspension containing approximately 8×10^2 cells/µl, at a flow rate of approximately 10 µl/min. Trypan blue was added to the

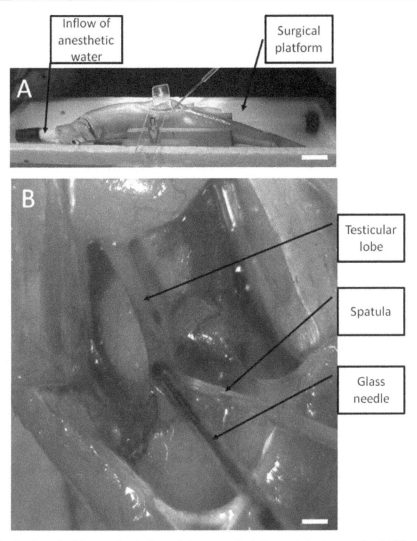

Fig. 3. Surgical transplantation of donor cells into recipient gonads. A) The recipient was placed on an operation platform and received a constant flow of oxygenated, cool water containing anesthetic through the gills. The transplantation of donor cell was performed through a small incision in the abdomen. B) Detail of the injection of the cell suspension (containing trypan blue as a marker) into the testis. Scale bars indicate 1cm (A) and 100 μm (B).

injection medium prior to transplantation to allow visualization of the cell suspension inside the needle and inside the gonad after injection. Repeated reentry and readjustment of the needle position were avoided to minimize tissue damage and leakage of cells. The abdominal incision was stitched with nylon surgical thread and topically treated with Isodine, and the fish were resuscitated in clean water.

Post-transplantation analysis of the fate of donor cells was performed first by fluorescent microscopy at 4 and 6 weeks after injection. For this purpose, the testes from three animals chosen randomly on each sampling were removed, washed in PBS (pH 8.2), macroscopically observed for the degree of dispersion of the cell suspension (Fig. 4), and immediately frozen in liquid nitrogen. Cryostat sections with a thickness of 10 μm were made from representative portions of these testes and observed under a fluorescent microscope for the presence of CFDA-SE-positive donor cells.

The fate of the donor cells was then examined by sperm counting and molecular (PCR) analysis between 6 and 12 months after the transplantation. On each occasion, 10-30 μl of sperm was collected from each of 20 transplanted males and 5 control males. Sperm was manually stripped by gentle abdominal pressure after careful removal of urine and wiping of the genital papilla. Ten microliters of sperm were then diluted 1,000 times with PBS and the density of spermatozoa was counted in a hemocytometer under a microscope. Some of the spreads of sperm used for counting were also observed under the fluorescent microscope for the detection of CFDA-SE labeled donor cells. Sperm samples collected at 6 and 8 months were subjected to PCR analysis for molecular detection of donor-derived cells.

DNA was extracted by the standard phenol:chloroform protocol (Annexure-II) and subjected to PCR analysis with *O. bonariensis*-specific primers (5´-CAG TGC AGG TCC AGC ATG GG-3´ and 5´-TGT TCC GCC TCA GTG CTT CAG-3´; amplicon size 386bp) that were designed Genetyx Ver 8.2.1. The PCR reactions were run in a Mastercycler EP Gradient S and consisted of an initial denaturation at 94°C for 3 min, 30 cycles of 94°C for 30 sec, 70°C for 30 sec and 72°C for 1 min, following elongation at 72°C for 5 min. PCR products were electrophoresed on an agarose gel (1%), stained with ethidium bromide, and photographed for posterior analysis.

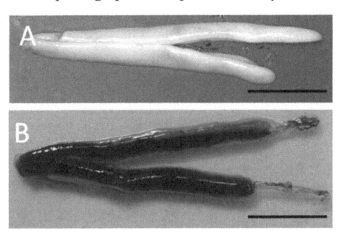

Fig. 4. Visualization of the dispersal of the cell suspension through the gonad after transplantation. A) Macroscopic appearance of the testes before germ cell transplantation. B) Appearance of the testes 4 weeks after the cell transplantation (note the diffusion of the marker trypan blue through all areas of the testis). Scale bars indicate 1 cm.

Once the sperm density of transplanted animals have returned to normal levels and the PCR analysis showed the presence of donor-derived sperm (e.g. after 6 months), artificial insemination was performed using the PCR-positive sperm to assess the viability of the *O. bonariensis* spermatozoa produced by the surrogate *O. hatcheri* fathers. About 10 µl of sperm from each of these males was used to fertilize a batch of *O. bonariensis* eggs, which were then incubated under standard conditions (flowing brackish water at 20°C) until hatching. The larvae from each cross were reared separately until sampling for PCR analysis of their genetic background at 8-10 days after hatching. The template DNA was extracted from each sample and subjected to PCR analysis using recipient's (*O. hatcheri*) specific markers (5′-ATG ATC AGC AGC TGA GCC CAC CTC C-3′ and 5′-TGT TCC GCC TCA GTG CTT CAG-3′; amplicon size 386bp) designed using Genetyx Ver 8.2.1. The PCR conditions were the same as described above. The genetic information on each individual was used to calculate the donor germline transmission rates.

The statistical significance of the differences in sperm production between groups was analyzed by one-way analysis of variance (ANOVA) followed by the Tukey's multiple comparison test using Graphpad Prism ver. 4.00. Data are presented as mean ± SD and differences between groups were considered as statistically significant at $P<0.05$.

The presence of donor-derived germ cells in recipient gonads, contrary to control (non-transplanted) gonads, was confirmed by tracking the CFDA-SE-labeled cells (Fig. 5 & 6A-F). The process of colonization of the recipient gonads by donor germ cells could be broadly divided into two phases. First, during the initial weeks, transplanted cells were randomly distributed throughout the seminiferous lobules (Fig. 5 & 6A) with only a small number reaching the blind end of the lobule (cortical region of the testis; Fig. 5 & 6B). This initial step of colonization, namely the settlement of stem spermatogonia in the cortical region of the testis, was evident in 2 out of the 3 fishes observed 4 weeks after transplantation. In the second phase, donor cells in the blind end of the lobules proliferated and formed a monolayer network along the cortical region of seminiferous lobules and establish a chain of cells (Fig. 5 & 6F). This stage was observed in 1 out of 3 fishes 6 weeks after transplantation.

Both the transplanted and the age-matched, non-transplanted males that were experimentally depleted of endogenous germ cells had low milt density at the beginning but the amount of sperm produced increased with time (Fig. 7). Recovery of the sperm count was faster in non-transplanted than in germ cell transplanted animals but both groups had similar values after 12 months. PCR testing of the sperm at 6 and 8 months after transplantation revealed the presence of *O. bonariensis* (donor)-derived spermatozoa in 4 out of 20 (20%) of the germ cell transplanted recipients (Fig. 8) and donor-derived spermatozoa were visually identified in the milt of these recipients by the presence of the fluorescent label (Fig. 9).

These four recipients with *O. bonariensis*-derived spermatozoa were subsequently used in progeny screening by artificial insemination of batches of eggs from *O. bonariensis* mothers. The crosses produced viable offspring with normal fertilization and hatching rates and which were estimated by PCR analysis to contain 1.2-13.3% pure *O. bonariensis* in addition to hybrids between the two species (Table 1, Fig. 10), suggesting that donor-derived germ cells could differentiate into fully functional spermatozoa in the allogeneic recipients. The contribution of donor-derived spermatozoa to the germline was estimated between 1.2 to 13.3% (Table 1).

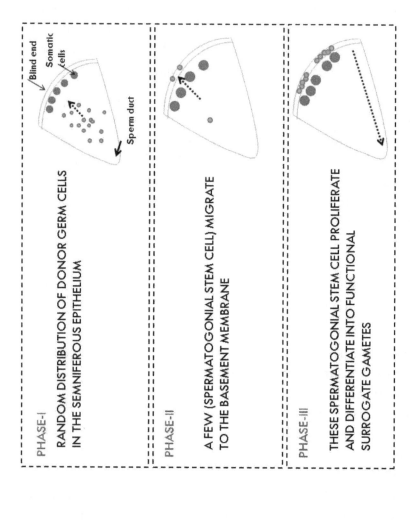

PHASE-I
RANDOM DISTRIBUTION OF DONOR GERM CELLS IN THE SEMNIFEROUS EPITHELIUM

PHASE-II
A FEW (SPERMATOGONIAL STEM CELL) MIGRATE TO THE BASEMENT MEMBRANE

PHASE-III
THESE SPERMATOGONIAL STEM CELL PROLIFERATE AND DIFFERENTIATE INTO FUNCTIONAL SURROGATE GAMETES

Fig. 5. Mechanism of donor germ cell migration, proliferation and differentiation inside the recipient gonads.

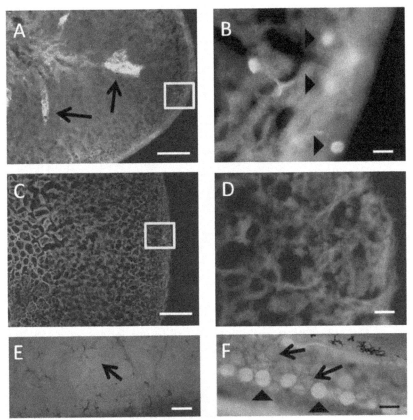

Fig. 6. Distribution of donor germ cells in recipient gonads 4-6 weeks after transplantation. A,B) Four weeks after transplantation, the donor germ cells were found distributed throughout the seminiferous lobules (arrows) and a small number, possibly spermatogonia, have reached the blind end of the seminiferous lobules (arrowheads; B is a high magnification of the box in A). C,D) Corresponding control sections from non-transplanted gonads (D is a high magnification of the box in C) showing the absence of fluorescent cells. E) Section of a non-transplanted control gonad showing a cyst of spermatogonia near the blind end of the seminiferous lobule at 6 weeks (arrow). F) Six weeks after transplantation, donor-derived (arrowheads) and endogenous (arrows) spermatogonia in recipient gonads have undergone proliferation along the blind end of the lobules. Scale bars indicate 100 μm (A, C) and 20 μm (B, D, E and F).

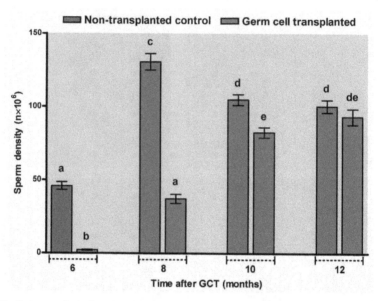

Fig. 7. Sperm density in donor cell transplanted recipients and non-transplanted (negative control) animals between 6 and 12 months after transplantation. Columns with different letters vary significantly (Tukey's multiple comparison test, $P<0.05$).

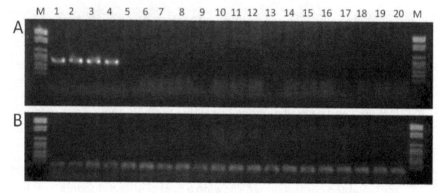

Fig. 8. PCR analysis of sperm from 20 donor cell transplanted recipients after 8 months. The analysis was conducted using *O. bonariensis*-specific marker (top panel) and β-actin (bottom panel). Donor-derived spermatozoa were detected in the sperm of four recipients (lanes 1-4, after a pre-screening at 6 months).

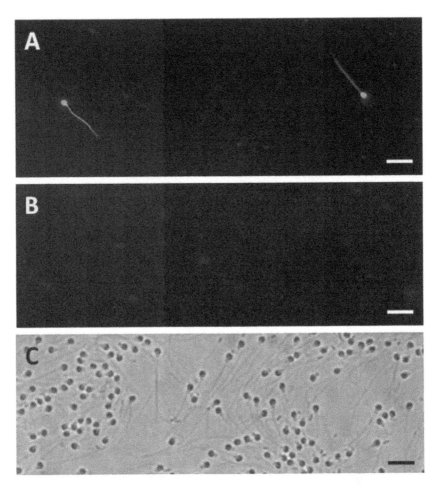

Fig. 9. Visualization of donor-derived spermatozoa in sperm collected from surrogate males 12 months after transplantation. Donor-derived cells are characterized by the fluorescence of the CFDA-SE marker as opposed to the non-fluorescent endogenous cells. Scale bar indicates 20 μm.

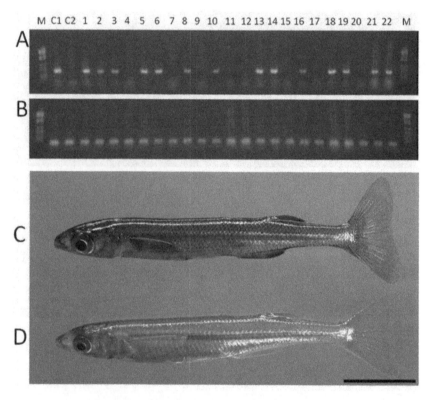

Fig. 10. PCR analysis of individual offspring from a cross between recipient male and an *O. bonariensis* female. The analysis was conducted using *O. hatcheri*-specific marker (A) and β-actin (B): the *O. hatcheri*-specific bands are detected only in hybrid individuals and absent in pure *O. bonariensis*. Control lanes include hybrids between an *O. hatcheri* male and an *O. bonariensis* female (C1) and a pure *O. bonariensis* (C2). Individual offspring are shown in lanes 1-22 (pure *O. bonariensis* are found in lanes 4, 7, 9, 11, 12, 15, 17, and 20, after a preliminary screening). A hybrid (C) and pure *O. bonariensis* (D) offspring derived from the cross of a surrogate *O. hatcheri* male transplanted with *O. bonariensis* (donor) germ cells and an *O. bonariensis* female. Scale bar indicates 1 cm.

It was observed that the donor germ cells harvested from sexually immature pejerrey *O. bonariensis* testes were successfully transplanted into partially-sterilized testes of Patagonian pejerrey *O. hatcheri* by surgical intervention. As previously reported in mammals, transplantation resulted in the recolonization of the seminiferous epithelium, resumption of spermatogenesis, and production of functional spermatozoa. Therefore, it was concluded 1) that repeated administration of Busulfan and rearing at a high temperature cause pronounced germ cell loss, 2) that testes treated in this way remain sexually-competent and able to support donor derived gametogenesis, as already noted previously, and 3) that germ cell transplantation into these gonads using simple injection techniques results in recolonization of the recipient gonads with the transplanted germ cell. Further, it was suggested that the simple but effective procedures demonstrated in this study could have immediate applicability in hatcheries and other seed production facilities, with great potential for the production of valuable species and the conservation (or restoration) of endangered species.

Irrespective of the differences in testicular structure between fish model used in this study and mammals (e.g. lobular and tubular testes, respectively), the process of recolonization of the recipient *O. hatcheri* gonads by the transplanted germ cells showed many similarities to that reported in germ cell transplanted mice. For instance, there was considerable interlobular variability in CFDA-SE-positive cell distribution during the first weeks after transfer but many labeled cells eventually settled along the blind end of the seminiferous lobules and began proliferating within weeks of transplantation. Given that only spermatogonial stem cells (type A spermatogonia) have the capability to migrate and settle at the basement membrane and resume the process of spermatogenesis, it was concluded that these cells in the cortical area were spermatogonial stem cells. This conclusion was also borne out by the fact that the recipients have been repeatedly stripped of sperm but continue to produce donor-derived spermatozoa. This is an indication that the transplanted cell population included cells with the potential for self-renewal and proliferation. Type A spermatogonia are able to undergo regular self-renewing divisions and maintain a pool of undifferentiated germ cells to support spermatogenesis, as opposed to the type B spermatogonia that rarely divide and go on differentiate into spermatocytes, spermatids, and finally spermatozoa. It is also noteworthy that the injected cells were able to migrate and settle at the germinal epithelium amidst the endogenous germ cells. It has been reported that transmembrane protein molecules present at the junctional complex located in the seminiferous tubules transduce signals, maintain cell polarity, and mediate germ cell migration. Although in this study, Majhi and colleagues did not examine the pathways involved in the migration of donor germ cells inside the recipient gonads, the results obtained in the study indicate that the pejerrey (donor) germ cells might sense and respond to molecules released from the blind end of the seminiferous lobules in Patagonian pejerrey (recipient) as both the species are closely related. A series of time-course observations prior to and after the 4-6 week post cell transplantation period also provide a more comprehensive view of the process of gonadal recolonization from donor-derived germ cell populations.

It was observed that donor-derived spermatozoa from surrogate fathers had similar functional properties as those of control ones in terms of fertilization and hatching rates, and could not detect any evidence of defective spermatogenesis. Yet, there have been cases of defective transcription of genetic information by donor cells during intertaxa germ cell transplantation, and these appear most prominently when the phylogenetic distance between the donor and recipient is large. In this study a congeneric fish model was used and the transplanted cells were probably immunologically compatible with the recipient's gonadal environment. Nevertheless, 16 of the 20 donor cell transplanted recipients ultimately did not develop donor-derived spermatozoa and the efficiency of recolonization of the recipient testes by donor cells decreased from approximately 2/3 of the individuals at 4 weeks, to 1/3 at 6 weeks, and finally 1/5 at >6 months. Thus, it cannot be ruled out entirely the possibility of immunological rejection and re-absorption of the transplanted cells. On the other hand, the variability observed at 4 weeks (e.g. animals with and without donor cells) might have originated, in part, from leakage of cell suspension from the seminiferous lobules at the time of injection. The variability in following weeks and months, on the other hand, may stem from poor control over the number of spermatogonial stem cells transferred into each recipient. Unfortunately, there are no specific biochemical or morphological markers currently available to identify spermatogonial stem cells from other germ cell types prior to transplantation. Alternatively, the number of germinal cell colonies recovered from seminiferous tubules in transplanted males can also be used as a proxy for the number of spermatogonial stem cells transferred into each recipient, but this strategy may not be a precise or timely tool for stem cell quantification, particularly when the recipients are not completely sterile. Therefore, development of genetic markers for stem cell identification might constitute a valuable tool for improving the efficiency of germ cell transplantation and for understanding the biology of stem cells.

Donor-derived germline transmission rate in the progeny ranging between 1.2 to 13.3% was recorded. In comparison, when primordial germ cells were transplanted into recipient embryos, their contribution to the germline was reported to be between 2 - 4%. In this context, the findings reported by Majhi and colleagues seem to support the view that transplanted spermatogonial cells have improved access to the germinal epithelium, where they are able to resume the spermatogenesis, if the recipients have been severely depleted of their endogenous germ cells prior to transplantation of exogenous germ cell. This may be particularly true with sexually competent animals as in this case. The proposed use of triploid sterile recipients, whose germ cells are unable to develop into fertile gametes, may seem to obviate the need for endogenous germ cell depletion. However, the long time required for producing and raising triploid animals to sexual maturity, and the need to develop the techniques for triploid induction in a species phylogenetically close to the target one may limit considerably the applicability of this approach for rare, endangered species. In this regard, the timely production of completely sterile fish, preferably by non-chemical means such as with the use of high temperature, may represent perhaps the only strategy available to save endangered species in situations when time to develop surrogate broodstock by germ cell transplantation has become a constraint.

In conclusion, the most striking result of this study was production of pure donor-derived progeny (up to 13.3%) in the span of months when donor germ cells were transplanted into the testes of sexually competent adult male recipients. This approach, which was validated for the first time in fish, offers a workable alternative to primordial germ cell or spermatogonia transplantation using embryos and/or young hatchlings, which requires sophisticated equipment, skills, and time to yield donor-derived gametes. Further, preferably low-tech, refinements in the proposed approach might render it even more useful for the timely rescue of rare or endangered fish species.

Table 1.   Results of artificial insemination of *O. bonariensis* eggs with sperm derived from surrogate *O. hatcheri* males transplanted with *O. bonariensis* germ cells.

Recipient males[*]	Number of eggs (n)	Fertilization (%)(n)	Hatching (%)(n)	Donor-derived germline transmission (%)(n)
#1	105	90.5 (95)	63.2 (60)	8.3 (5)
#2	75	100.0 (75)	80.0 (60)	13.3 (8)
#3	96	93.7 (90)	88.8 (80)	1.2 (1)
#4	100	95.0 (95)	97.9 (93)	3.2 (3)
Control	175	100.0 (175)	85.7 (150)	NA

[*] PCR-positive recipients that produced donor-derived gametes 6 months after the GCT.

CHAPTER-4

■——■

SURROGATE BROODSTOCK DEVELOPMENT BY NON-SURGICAL APPROACH

Various assisted reproductive technologies have been devised to efficiently produce functional gametes and offspring from endangered species and commercially important animals that are difficult to breed in captivity. These approaches include cryopreservation of gametes and embryos, induction of multiple ovulations, embryo transfer, *in vitro* gametogenesis, nuclear transfer, and germ cell transplantation, among others. Germ cell transplantation provides also a unique system for studying the cellular and molecular events that regulate the sequential steps of gonadogenesis and gametogenesis. There is particular interest in developing efficient methods of germ cell transplantation for surrogate broodstock development in fish due to the growing concern with dwindling fisheries stocks and loss of species/ genetic biodiversity due to over exploitation, environmental degradation and increasing effects of climate change (Fig. 1). The success of surrogate broodstock development largely depends on the availability of recipients that are completely or partially devoid of endogenous germ cells. The recipient gonads must be also genetically compatible with the donor species but most recipients seem to present little or no rejection to the transplanted cells even if they are from relatively unrelated donors. This fact makes possible to use domesticated strains and/or prolific species as recipients in germ cell transplantation.

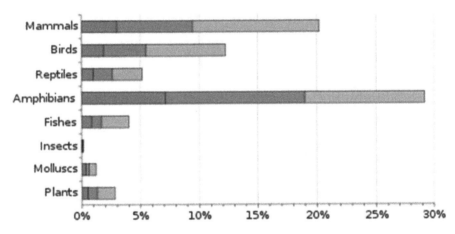

Fig. 1. The percentage of species in several groups which are listed as critically endangered, endangered, or vulnerable on the IUCN Red List.

Several options for eradication of endogenous germ cells in recipients have been tested in mammals such as treatment with cytotoxic drugs like Busulfan, irradiation, cold ischemia and hyperthermic treatment (see Chapter-2). Two types of recipients have been experimentally tested for germ cell transplantation in fish. Triploid animals have been used by the Japanese scientists for production of donor-derived gametes in salmonids taking advantage of the fact that they are generally, though not always, sterile. However, this strategy requires the long-term rearing of recipient animals until adult size as triploids can be produced only by manipulation of genetic events during or shortly after fertilization. An alternative is the use of recipient fish which are depleted of endogenous germ cells by chemical and heat-cytoablative treatments. One advantage of this approach is that, when applied to sexually competent adult, it obviates long-term rearing of hosts and allows surrogate generation of gametes within a relatively short time after transplantation of donor cells. For instance, Majhi and colleagues in a previous report demonstrated that recipients prepared with such strategy and transplanted with donor germ cells produced donor-derived functional gametes within 6 months, with germline transmission rates of 1.2-13.3%. In that study, the recipients were prepared by rearing at a temperature of 25°C and by administration of two doses of Busulfan (40 mg/kg BW) at 4 weeks intervals. However, it was observed that many females developed ulcerations shortly after Busulfan treatment and suffered increased mortality not observed in males.

The method of transplantation also has variants such as microinjection of germ cells in the blastodisc of blastula stage embryos, into the coelomic cavity of hatchlings, and directly into gonads of adults by surgical or non-surgical (intra-papillar) intervention. Regardless of their advantages and disadvantages, germ cell transplantation by all methods and at all developmental stages has led to production of donor-derived functional gametes. However, there are obvious differences in the level of skills and equipment required to perform germ cell transplantation by each of these methods, and some may be inapplicable in remote areas of the world where conservation efforts are probably more necessary. More importantly, they entail a fundamental difference in the time needed for production of surrogate gametes as previously mentioned, particularly for the comparison between germ cell transplantation in embryos/hatchlings and in sexually competent adults.

Germ cell transplantation into the ovary of adult females has never been explored, thereby posing a constraint to the production of female gametes in situations where both sexes need equal attention for conservation and propagation. In this context, germ cell transplantation procedure of Majhi *et al.*, 2009 and Lacerda *et al.*, 2010 were re-evaluated in this study for sexually competent adult fish in order to 1) optimize the thermo-chemical treatments for enhancing germ cell niche availability for germ cell transplantation while minimizing the occurrence of pathologies in recipients and 2) to examine the suitability of thermo-chemically sterilized gonads to support the colonization, proliferation and differentiation of foreign germ cells transplanted by non-surgical, intra-papillar intervention. More importantly, germ cell transplantation in female recipients was performed first time, in addition to males. Thus, using the congeneric recipient Patagonian pejerrey *Odontesthes hatcheri* and donor species pejerrey *Odontesthes bonariensis,* functional viability of gametes from surrogate parents of both sexes produced by intra-papillar transplantation in adult fish was performed. The simplified germ cell transplantation approach described in this chapter ultimately leads to surrogate production of eggs and sperm in considerably short time, therefore being suitable for timely *in vivo* propagation of genetic resources.

Methods for surrogate gametes and offspring production

One year old Patagonian pejerrey *Odontesthes hatcheri* (mean body weight ± SD of 37.35 ± 16.8 g for males and 33.1 ± 13.6 g for females) were used as recipients. Fish were stocked in 200 L tanks at a density of 7.5 kg of fish per m^3 and reared in flowing brackish water (0.2-0.5% NaCl) under a constant light cycle (15L9D). The animals were acclimated for two weeks at 17°C prior to the thermo-chemical treatments. The germ cell donors were 4-5 months old specimens of the congeneric species pejerrey *O. bonariensis* (mean body weight ± SD of 0.91 ± 0.1 g for males and 0.8 ± 0.03 g for females). Donors were stocked in 200 L tanks at a density of 1.0 kg of fish per m^3 and reared in flowing brackish water (0.2-0.5% NaCl) under a constant light cycle (15L9D) at temperature (25°C) until use. Both groups of animals were fed pelleted commercial diet four times per day to satiation.

The endogenous germ cell of recipient fish was depleted by combination of thermo (26°C)-chemical (Busulfan) treatments. The number of Busulfan injections was increased from two to four to accelerate germ cell loss. On the other hand, dosage of Busulfan for females was decreased (30 mg/kg body weight; B30) compared to males (40 mg/kg body weight; B40) because the highest dosage appeared to be toxic to females. Each of the experimental groups had 80 fish of each sex. Females and males in the control group (B0; 30 fish of each sex) received only the vehicle DMSO at 1% Kg BW. The four injections were administered at the start (day 0) and at the 2nd, 4th, and 6th weeks. For checking the permanency of the germ cell loss, animals were transferred from 26°C to 17°C at 8 weeks and reared for an additional period of 16 weeks (recovery period).

For histological observations on the process of germ cell loss and the permanency of the germ cell deficiency, 10 males and 10 females were randomly sampled from each group at the 4th, 8th, and 24th weeks (end of the recovery period). Animals were killed by an overdose of anesthesia and their body weight was recorded. The gonads were dissected, macroscopically examined, photographed using a digital camera, and weighed to the nearest 0.01g. The middle portion of the right and left gonads from each fish were then immersed in Bouin's fixative for 24 hours and preserved in 70% ethanol. Gonads were processed for light microscopical examination following routine histological procedures up to preparation of thin (5 μm) sections and staining with hematoxylin-eosin. About 40-50 serial histological sections from each fish were examined under a microscope at magnifications between 10-60X. The degree of histological degeneration and germ cell loss of each specimen was classified following the criteria described in Chapter-2.

The degree of GC loss in *O. hatcheri* gonads was monitored by Real Time RT-PCR analysis of *vasa* mRNA expression according to the protocols of Majhi *et al.* 2008. Briefly, samples from the middle part of the gonads were collected from all groups at the same time of the histological samples and stored in RNA *later* at -80°C until further processing. RNA was extracted using Trizol according to manufacturer's protocol. cDNA was synthesized using oligodT primers and Superscript reverse transcriptase. Primers for Real-Time RT-PCR (forward: 5´-CCTGGAAGCCAGGAAGTTTTC-3´; reverse 5´-GGTGCTGACCCCACCATAGA-3´) were designed using Primer Express (ver. 2.0). The Real Time PCRs were run in an

ABI PRISM 7300 using *Power*SYBR® Green PCR Master Mix in a total volume of 15 ul which included 25 ng of first strand cDNA and 5 pmol of each primer. β-actin (forward: 5´-CTCTGGTCGTACCACTGGTATCG-3´; reverse: 5´-GCAGAGCGTAGCCTTCATAGATG-3´) was analyzed as an endogenous control. Quantification was performed using the standard curve method with 4 points and the ABI Prism 7300 Sequence Detection Software (v.1.2).

Donor cells from male and female *O. bonariensis* were isolated based on the protocol described by Majhi *et al.* 2009. Briefly, fish were sacrificed by anesthetic overdose and the gonads were excised and rinsed in phosphate-buffered saline. The gonadal tissue was finely minced and incubated in a dissociating solution containing 0.5% Trypsin, 5% Fetal Bovine Serum, and 1mM Ca^{2+} in PBS for 2 hr at 22°C. The dispersed gonadal cells were sieved through a nylon screen (mesh size 50 µm) to eliminate non-dissociated cell clumps, suspended in a discontinuous Percoll gradient, and centrifuged at 200×g for 20 min at 20°C. The phase containing predominantly gonial cells (spermatogonia or oogonia) was harvested and the cells were rinsed and subjected to a cell viability test by the Trypan blue (0.4% w/v) exclusion assay (Fig. 2). The PKH 26 Cell Linker kit was then used to label the cells to determine their localization inside the recipient gonads. The cells were labeled with the dye at the concentration of 8 µl per mL for 10 min at room temperature (Fig. 3) and the staining was stopped by addition of an equal volume of heat-inactivated fetal bovine serum. Labeled cells were rinsed three times to remove unincorporated dye, suspended in Dulbecco Modified Eagle Medium with 10% fetal bovine serum, and stored on ice until transplantation.

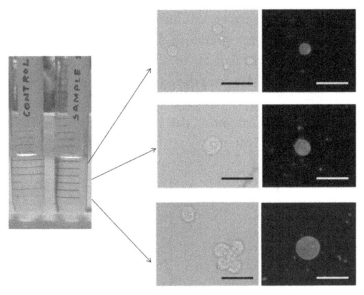

Fig. 2. Isolation of cells by discontinuous Percoll gradient. The bottom phase containing predominantly gonial cells (spermatogonia or oogonia) was harvested and labeled with PKH dye for transplantation.

On termination of heat-Busulfan treatments at 8 weeks, the water temperature of the experimental tank was gradually decreased (1-2°C/day) to pretreatment condition. Once the rearing temperature reached 17°C, germ cell transplantation was performed in 50 B40 males and 50 B30 females. Briefly, the fish were anesthetized in 100 ppm Phenoxyethanol and held upside down onto an operation platform under a microscope where they received a constant flow of oxygenated, cool water containing 100 ppm of the anesthetic through the gills. To prevent desiccation, the surface of the fish was moisturized during the entire donor cell transplantation procedure, which took about 8-10 min per fish on average. A micro syringe and fine glass needle were used to inject the cell suspension into the gonads through the genital papilla (Fig. 4). Each individual was injected with 50 µl of cell suspension containing approximately 7.5×10^4/ µl, at a flow rate of approximately 10 µl/min. Trypan blue was added to the injection medium to allow visualization of the cell suspension inside the needle and leakage during/after transplantation. The genital opening of each fish was topically treated with 10% Isodine after the procedure and the fish were resuscitated in clean water.

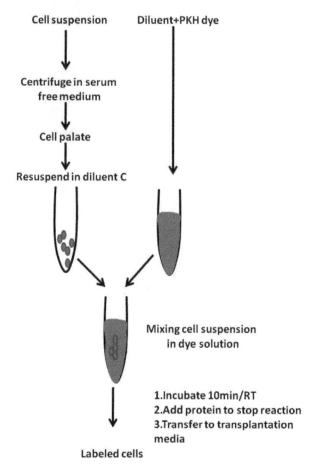

Fig. 3. Protocol for labeling of donor germ cell by PKH-26 dye.

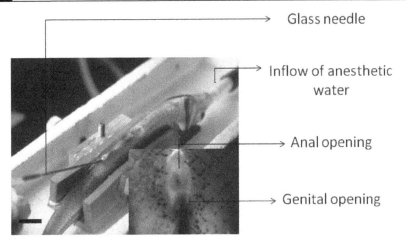

Glass needle

Inflow of anesthetic water

Anal opening

Genital opening

Fig. 4. Intra-papillar transplantation of donor cells into recipient gonads. The recipients were placed onto an operation platform and received a constant flux of aerated anesthetic water through the gills during the procedure. The medium containing the donor cells was visualized by addition of Trypan blue during injection through the genital papilla (inset shows magnified view of injection). Scale bar indicates 1 cm.

The fate of donor cells was assessed preliminarily by microscopical observation of the PKH 26-labeled cells in gonadal sections at 2, 4, 6, 8, and 24 weeks after injection. For this purpose, gonads were excised from 3-5 transplanted animals for each sampling, washed in PBS, fixed in 4% formaldehyde overnight at 4°C, immersed in 15% sucrose for 2-3 hrs, embedded in O.C.T. compound, frozen using dry ice and stored at -80°C until actual sectioning. Cryostat sections with a thickness of 10 μm were made from representative portions of the gonads, air dried for 45-60 min at room temperature, coverslip mounted using 1-2 drops of scotch instant glue, and observed under a fluorescent microscope. Control sections were prepared using the gonads of animals not subjected to the transplantation procedure. Images were captured using a Pixera digital camera and software (Viewfinder/Studio).

The presence of donor-derived gametes in the donor cell transplanted recipients was examined by molecular (PCR) analysis 7 months after transplantation. For the analysis of males, 10-30 µl of sperm was manually stripped by gentle abdominal pressure in arcas around the genital papilla. In females, 30-50 eggs were collected from each female by cannulation. DNA from sperm and eggs was extracted by the standard phenol:chloroform protocol and subjected to PCR analysis with *O.bonariensis*-specific primers (forward: 5′-CAGTGCAGGTCCAGCATGGG-3′ and reverse: 5′-TGTTCCGCCTCAGTGCTTCAG-3′; amplicon size 386bp) and *O. hatcheri*-specific primers (forward: 5′-ATGATCAGCAGCTGAGCCCACCTCC-3′ and reverse: 5′-TGTTCCGCCTCAGTGCTTCAG-3′; amplicon size 386bp) that were designed based on the sequence of the first intron of the *amha* genes of these species using Genetyx Ver 8.2.1. Primers for β-actin indicated in *vasa* gene expression analysis were used as positive controls. The PCR reactions were run in a Mastercycler EP Gradient S and consisted of an initial denaturation at 94°C for 3 min, 30 cycles of 94°C for 30 sec, 70°C for 30 sec, and 72°C for 1 min, following elongation at 72°C for 5 min. PCR products were electrophoresed on an agarose gel (1%), stained with ethidium bromide, and photographed for later analysis.

Gametes taken from the surrogate parents at 7 and 11 months after the cell transplantation were used in artificial insemination together with eggs and sperm from pure *O. bonariensis* mothers and fathers. The fertilized eggs obtained from each cross were incubated at 17°C for subsequent analysis of fertilization, hatching and germline transmission rates (%). The template

DNA used in the germline transmission rate analysis was extracted from each progeny within 7-10 days after hatching and subjected to PCR analysis using both *O. bonariensis*-specific and *O. hatcheri*-specific sequences and conditions indicated above.

Measured parameters were compared among the treatments by one-way analysis of variance (ANOVA) followed by the Tukey test whereas donor-derived germline transmission rates at 7 and 11 months after germ cell transplantation were compared by the Fisher's exact test. Both statistical analyzes were performed with GraphPad Prism ver. 6.00. Data are presented as mean ± SE and differences between groups were considered as statistically significant at $P<0.05$.

The survival rate of both sexes was high and the minimum rate (95%) was recorded for Busulfan-treated females (B30). There was a significant reduction in body weight between the 4th and 8th weeks for both treated females (B30) and males (B40), as the fishes apparently reduced food intake compared to the untreated controls. This was particularly noticeable after administration of the 3rd dose on the 4th week (Fig. 5). However, the fishes regained weight once the treatment was discontinued and they were returned to the normal temperature (17°C). External pathologies such as the ulcerations observed in previous studies were not detected in this study.

The gonado-somatic index (GSI) of males, including the controls (B0), decreased significantly between 0 and 4 weeks but only those in the group B40 continued decreasing further up to 8 weeks (Fig. 6). Microscopic examination of the gonads, on the other hand, revealed that all of the 10 control animals by 4 weeks

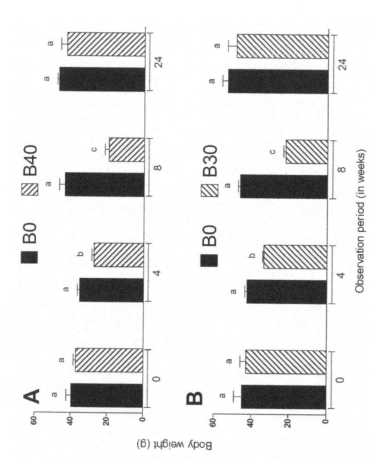

Fig. 5. Changes in mean body weight of males (A) and females (B) subjected to heat (26°C) and Busulfan treatments (B0: Busulfan 0 mg/kg, controls; B30: 30 mg/kg, only females; B40: 40 mg/kg, only males) between 0 and 8 weeks and of Busulfan-treated animals after recovery for 16 weeks at 17°C (total 24 weeks). Columns with different letters vary significantly (ANOVA - Tukey test, $p<0.05$).

and 5 out of 10 animals at 8 weeks had active spermatogenesis (Fig. 7A-B; Table 1 and 2). The remaining 5 controls had relatively shrunk gonads and only cysts of spermatogonia. In contrast to the controls, B40 males at 4 weeks (total of 2 doses) had testes showing from absence of any spermatogenic stages beyond spermatogonia to the complete disappearance of all germ cells including spermatogonia. In addition, some testes presented conspicuous cysts of abnormal cells with variable size and staining properties ranging from dense basophilic (hematoxylin) to acidophilic (eosin) bodies (Fig. 2C-D). At 8 weeks (total of 3 doses), the degree of germinal degeneration was far more severe and 9 out of 10 animals were found to be consistently devoid of germ cells in all histological sections examined (Fig. 7E-F). However, after 16 weeks of recovery at 17°C, presumed sterility was confirmed only in 4 out of 10 animals (Fig. 7G-H; Table 2). The histological findings were corroborated by the Real Time RT-PCR analysis, which showed that *vasa* transcript levels were significantly lower in B40 males at 4 and 8 weeks compared to B0 controls ($P<0.05$; Fig. 8A). Upon recovery for 16 weeks at 17°C, the transcript levels of B40 males remained low whereas those in B0 rebounded to pre-treatment levels.

The GSI of B0 females did not vary significantly between 0 and 8 weeks whereas that of B30 showed significant decreases (Fig. 6B). Control females did not show any marked histological changes throughout the study (Fig. 9A-B) but 4 out of 10 individuals at 8 weeks showed light atrophy of the ovigerous lamellae, fewer oocytes, and presence of degenerating oogonia (Table 2). In the B30 group, a similar histological characteristic

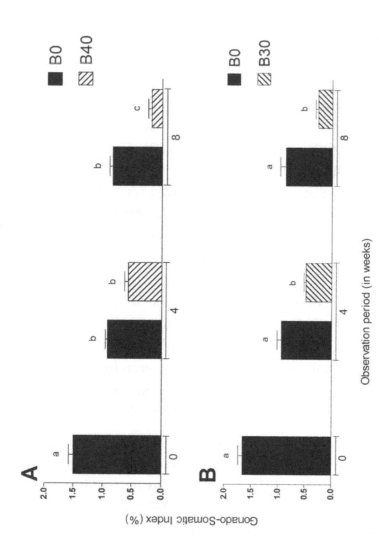

Fig. 6. Changes in the gonado-somatic index of males and females subjected to heat (26°C) and Busulfan treatments (B0: Busulfan 0 mg/kg, B30: 30 mg/kg, B40: 40 mg/kg) between 0 and 8 weeks. Columns with different letters vary significantly (ANOVA - Tukey test, $P<0.05$).

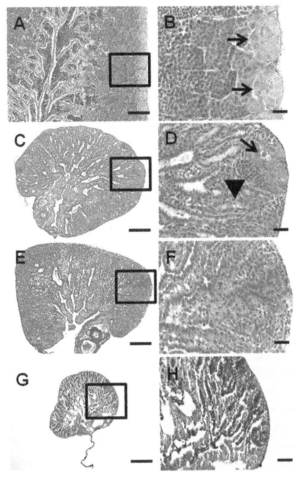

Fig. 7. Histological changes in the testes of males subjected to heat (26°C) and Busulfan treatment (40 mg/kg). Panels on the right are high magnifications of insets in the left panels. A,B) Start of treatment (week 0; arrows indicate large cysts of spermatogonia in the blind end of the spermatogenic lobules; note also active spermatogenesis within the lobules). C,D) 4 weeks (arrow indicates an abnormal spermatogonium). E,F) 8 weeks (note the absence of spermatogonia). G,H) 24 weeks (note complete absence of GCs even after recovery for 16 weeks at 17°C). Scale bars indicate 100 μm (A, C, E, and G) and 20 μm (B, D, F and H).

was observed in 5 individuals as early as 4 weeks. Furthermore, 5 out of 10 and 2 out of 10 females sampled at 4 and 8 weeks, respectively, had remarkable reduction in oogonial population as well as deposition of yellowish-brown pigments, an indication of the occurrence of macrophage phagocytic activity (Fig. 9C-D). The remaining 8 females from group B30 that were examined at 8 weeks were completely devoid of oogonia in all sections examined (Fig. 9E-F; Table 2). Following 16 weeks of recovery at 17°C, ovigerous lamellae were disorganized and oogonia were still missing altogether in 4 out of 10 females in this group (Fig. 9G-H). The *vasa* transcript levels were significantly lower at 4 and 8 weeks in B30 females compared to the respective controls (P<0.05; Fig. 8B). Upon recovery for 16 weeks, transcript levels recovered in the B0 but remain low in B30 group.

Table 1. **Histological criteria for classification of gonadal integrity/ degeneration in adult Patagonian pejerrey *Odontesthes hatcheri*.**

Males	Class	Females
Cysts of spermatogonia and other spermatogenic stages development	I	Cysts of oogonia interspersed with oocytes at various stages of
Only cysts of spermatogonia; efferent ducts may or not contain residual spermatozoa	II	Light hypertrophy of the ovigerous lamellae; oocytes are few and atretic
Cysts of spermatogonia are few and small	III	Few if any oocytes and greatly reduced number of oogonia
Absence of spermatogonia and any other germ cells	IV	Absence of oogonia and any other germ cells

Table 2. Frequency of individuals per category of histological appearance of the gonads in adult Patagonian pejerrey (*Odontesthes hatcheri*) subjected to heat (26°C) and Busulfan treatments (B0: Busulfan 0 mg/kg, controls; B30: 30 mg/kg, only females; B40: 40 mg/kg, only males) for germ cell depletion. Histological categories are described in Table 1.

Treatments	Observation period (weeks)	Sex	Number of fish			
			Histological category			
			I	II	III	IV
B0	4	♂	10	–	–	–
		♀	10	–	–	–
	8	♂	5	5	–	–
		♀	6	4	–	–
	24 (Recovery)	♂	10	–	–	–
		♀	10	–	–	–
B30	4	♂	–	5	5	–
	8	♀	–	–	2	8
	24 (Recovery)	♂	5	1	–	4
B40	4	♂	–	–	7	3
	8	♂	–	–	1	9
	24 (Recovery)	♂	5	–	1	4

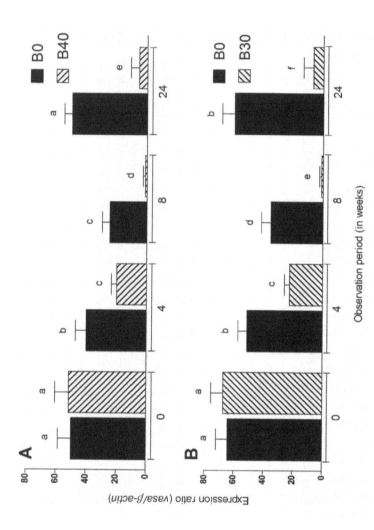

Fig. 8. Changes in *vasa* gene transcript levels in the control (A; B0: Busulfan 0 mg/kg) and treatment (B; B40: Busulfan 40 mg/kg) subjected to constant heat (26°C) between 0 and 8 weeks and recovery of treated fish after 16 weeks at 17°C (total 24 weeks). Columns with different letters vary significantly (ANOVA - Tukey test, *P*<0.05).

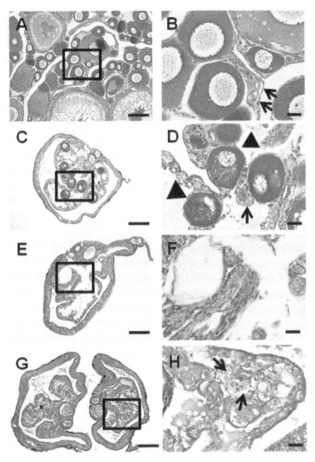

Fig. 9. Histological changes in the ovaries of females subjected to heat (26°C) and Busulfan treatment (30 mg/kg). Panels on the right are high magnifications of insets in the left panels. A,B) Start of the experiment (week 0; arrows show a prominent cyst of oogonia; note also oocytes at various stages indicating active oogenesis). C,D) 4 weeks (note the absence of prominent cysts of oogonia, degenerating perinucleolar oocytes (arrowhead), and macrophage phagocytic activity indicated by deposition of yellowish-brown pigments (arrow). E,F) 8 weeks (note the absence of oogonia and other types of germ cells). G,H) 24 weeks (note the absence of germ cells, disorganized ovigerous lamellae (arrows), and hypertrophy of the tunica albuginea). Scale bars indicate 100 μm (A, C, E, and G) and 20 μm (B, D, F and H).

In males, the transplanted donor-derived germ cells were found randomly distributed throughout the spermatogenic lobules in all 5 recipient males examined 2 weeks after transplantation. At 4 and 6 weeks, a small number of donor cells (presumably spermatogonia stem cells) had reached the blind end of the lobules (cortical region of the testis; Fig. 10C-D) in all animals examined. The transplanted germ cells then proliferated and formed conspicuous cysts visible along the cortical region of the testis; this stage was observed in 4 out of 5 animals 8 weeks after transplantation (Fig. 10F). After 6 months, donor germ cell cysts were undergoing differentiation and displacement towards the efferent ducts in 2 out of 5 recipients (Fig. 10G-I). Sperm could be collected by gentle abdominal stripping in 20 out of 23 recipients examined 7 months after transplantation; the remaining 3 males were considered to be sterile based on histological analysis. Three of the spermiating animals were accidentally lost before sufficient sperm could be sampled for PCR analysis. PCR analysis of the remaining 17 germ cell-transplanted recipients revealed the presence of donor-derived cells together with endogenous cells in 17% (3 out of 17) (Fig. 11).

The transplanted cells in females were randomly distributed throughout the ovarian lumen and the surface of the ovarian lamella during the first 2 weeks after germ cell transplantation (Fig. 12A-B). At 4 and 6 weeks, a small number of donor germ cells, presumably oogonia, formed clusters in the ovaries of 3 out of 5 recipients (Fig. 12C-D). Larger clusters of oogonia were observed at 8 weeks and 6 months after germ cell transplantation the transplanted cells had differentiated into mature oocytes in

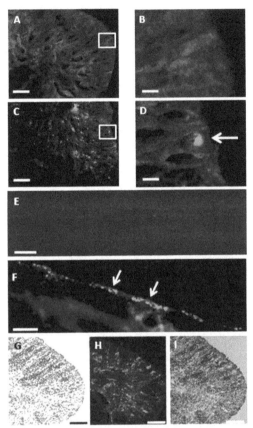

Fig. 10. Fate of PKH-26-labeled donor germ cells in recipient testes between 4 and 24 weeks after germ cell transplantation. A,B) Cryostat section of a non-transplanted, control testis at 4 weeks showing the approximate location of the blind end of the spermatogenic lobules (B is a high magnification of the box in A). C,D) Cryostat section of a transplanted testis at 4 weeks showing the presence of transplanted germ cells at the blind end of the spermatogenic lobules (arrow; D is a high magnification of the box in C). E) Whole-mount preparation of a non-transplanted control testis at 8 weeks. F) Whole-mount preparation of a transplanted testis at 8 weeks showing the presence of donor-derived germ cells (arrows) along the length of the gonad. G-I) Cryostat (G), corresponding HE (H) and merged (I) sections of a transplanted testis at 6 months showing differentiation of the donor-derived cells along the spermatogenic lobules towards the efferent ducts. Scale bars indicate 100 μm (A, C, E, F, G, H and I) and 20 μm (B and D).

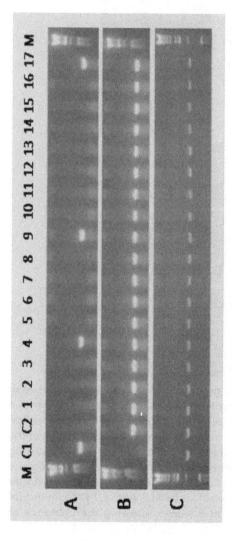

Fig. 11. PCR analysis of sperm from 17 male recipients 7 months after germ cell transplantation. The primers used in the analysis were based on an *O. bonariensis*-specific sequence (A), an *O. hatcheri*-specific sequence (B), and β-actin (C) as a template control. Control lanes include pure *O. bonariensis* (C1) and *O. hatcheri* (C2) sperm. Donor-derived *O. bonariensis* spermatozoa were detected in the sperm of three surrogate *O. hatcheri* recipients shown in lanes 4, 9, and 17.

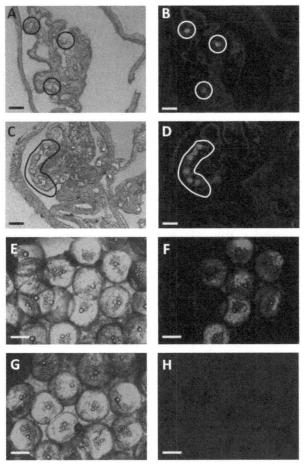

Fig. 12. Fate of PKH 26-labeled donor germ cells in recipient ovaries between 2 and 24 weeks after germ cell transplantation. A,B) Cryostat section of a transplanted ovary at 2 weeks showing donor germ cells randomly distributed throughout the ovarian lamellae (circles; B is a fluorescent view of bright field A). C,D) Cryostat sections of a transplanted ovary at 4 weeks showing the donor germ cells (presumably oogonia) forming aggregations (highlighted; D is a fluorescent view of bright field C). E,F) Whole-mount preparation of oocytes from a transplanted female at 6 months showing the presence of fully differentiated donor-derived oocytes (characterized by retention of fluorescent label). G,H) Whole-mount preparation of oocytes from a non-transplanted control female. Scale bars indicate 20 μm (A, B, C and D) and 100 μm (E, F, G and H).

1 out of 5 females examined (Fig. 12E-F). At 7 months, ripe eggs were collected from recipients by intra-ovarian cannulation and the PCR analysis showed the presence of donor-derived cells along with endogenous cells in 5% (1 out of 20) ovulating recipients (Fig. 13).

The germ cell transplantation recipients found by PCR analysis to produce *O. bonariensis*-derived spermatozoa and eggs were then subjected to progeny testing by artificial insemination using eggs and sperm from pure *O. bonariensis* females and males 7 and 11 months after germ cell transplantation. Such crosses produced viable offspring with normal fertilization and hatching rates compared to control animals. The crosses between surrogate males and *O. bonariensis* females yielded between 12.6 and 39.7% pure *O. bonariensis* in addition to hybrids between the two species; the donor-derived germline transmission rates 11 months after germ cell transplantation did not show a clear trend of increase or decrease compared to those at 7 months (Table 3; Figure 14A-G). The cross between the surrogate female and an *O. bonariensis* male yielded 52.2% and 39.7% pure *O. bonariensis* at 7 and 11 months, respectively, in addition to hybrids between the two species (Table 4).

The efficiency of germ cell transplantation generally improves when the recipient gonads are devoid of endogenous germ cells because of increased stem cell niche availability and accessibility to implanted cells. Multiple injections of Busulfan or repeated exposure to Gamma-ray radiation have been often used in adult mammalian and avian species to ablate the endogenous germ cells prior to germ cell transplantation. In fish, injection of a single dose of Busulfan or two dosages in association with elevated

Table 3. Results of artificial insemination of eggs derived from pure pejerrey females with sperm from three surrogate Patagonian pejerrey males (transplanted with pejerrey donor germ cells) and a control pure pejerrey male. Donor-derived germline transmission rates for each surrogate male were determined at 7 and 11 months after germ cell transplantation; asterisks after the donor-derived germline transmission rate for 11 months indicate significant difference (Fisher's exact test) from the rate for the same male at 7 months. NA: not applicable

Male	Time after GCT germ cell transplantation (months)	Number of eggs (n)	Fertilization rate (n; %)	Hatching rate (n; %)	Donor-derived germline transmission rate(n; %)
#1	7	160	158(98.7)	150(94.9)	19(12.6)
	11	140	131(93.5)	122(93.1)	30(24.6)*
#2	7	155	147(94.8)	141(95.9)	56(39.7)
	11	170	162(95.2)	150(92.5)	44(29.3)
#3	7	130	125(96.1)	118(94.4)	38(32.2)
	11	155	146(94.1)	135(92.4)	28(20.7)*
Control	7	140	132(94.2)	125(94.6)	NA
	11	165	159(96.3)	148(93.0)	NA

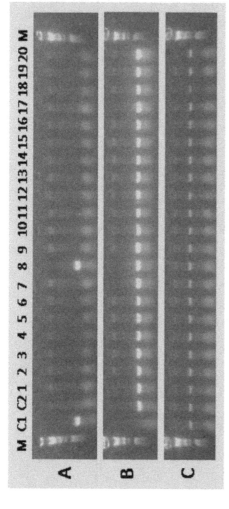

Fig. 13. PCR analysis of eggs from 20 female recipients 7 months after germ cell transplantation. The primers used in the analysis were based on an *O. bonariensis*-specific sequence (A), an *O. hatcheri*-specific sequence (B), and β-actin (C) as a template control. Control lanes include pure *O. bonariensis* (C1) and an *O. hatcheri* (C2) eggs. Donor-derived *O. bonariensis* eggs were detected in one surrogate *O. hatcheri* recipient shown in lane 8.

Table 4. Results of artificial insemination of eggs derived from a surrogate Patagonian pejerrey female (transplanted with pejerrey donor germ cells) and from a control pure pejerrey female with sperm from a pure pejerrey male. Donor-derived germline transmission rates determined for the same surrogate female at 7 and 11 months after germ cell transplantation were not significantly different (Fisher's exact test). NA: not applicable.

Female	Time after germ cell transplantation (months)	Number of eggs (n)	Fertilization rate (n; %)	Hatching rate (n; %)	Donor-derived germline transmission rate(n; %)
#1	7	108	98(90.7)	90(91.8)	47(52.2)
	11	120	108(90)	98(90.7)	39(39.7)
Control	7	105	100(95.2)	95(95.0)	NA
	11	130	123(94.6)	115(93.4)	NA

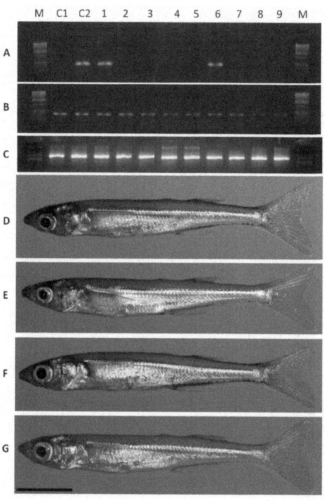

Fig. 14. PCR analysis of offspring from the crosses between surrogate *O. hatcheri* and pure *O. bonariensis* individuals. A-C) Results for the offspring of surrogate male. The primers used in the analysis were based on an *O. hatcheri*-specific sequence (A), an *O. bonariensis*-specific sequence (B), and β-actin (C) as a template control. Control lanes include a pure *O. bonariensis* (C1) and a hybrid from an *O. hatcheri* male and an *O. bonariensis* female (C2). Individuals shown in lanes 1 and 6 are hybrids and those shown in lanes 2-5 and 7-9 are pure, donor-derived *O. bonariensis* (D-G). Scale bar indicates 2 cm.

water temperature led to considerable depletion of germ cells in sexually mature gonads, but complete sterility has never been achieved. Previous study with sub-adult Patagonian pejerrey (*O. hatcheri*) showed that 2 doses of 40 mg Busulfan/kg BW and rearing at 25°C for 8 weeks led to marked depletion of germ cells. However, the treatment did not result in complete sterility and caused development of ulcerations and increased mortality in females. In this study efficiency of further (four) doses of Busulfan and increased temperature (26°C) to promote endogenous germ cell loss was tested. Lower dosage for females (30 mg Busulfan/kg BW) compared to males (40 mg/kg BW) in an attempt to prevent the appearance of pathologies was also tested. The treatments resulted in high percentages of fishes histologically classified as "sterile" at the end of the thermo-chemical treatment (8 weeks, 80-90% of the animals) and at the end of the recovery period at 17°C (6 months, 40% of the animals). These results compare well to none in previous study, although the results of gamete collection at 7 months after transplantation of donor cells clearly indicate that only a few fish, about 13% of the males and no females, were completely sterile. Moreover, although depressed food intake, leading to temporarily decreased body weight in both sexes, no other acute side effects and/or pathologies like ulcerations or deformities were observed. These observations confirm the suitability of multiple injection of Busulfan at concentrations of 30 mg/kg BW and 40 mg/kg BW for adult Patagonia pejerrey females and males, respectively, during recipient preparation for surrogate broodstock development.

Since the previous report on the feasibility of surgical intra-gonadal germ cell transplantation in adult fish, germ cell transplantation in sexually competent animals has been performed also in the tilapia model. Lacerda's group was the first to produce donor fish sperm in surrogate fathers using intra-papillar transplantation of donor cells but the usefulness of this simple procedure was not been asserted in females. This study confirmed the usefulness of intra-papillar germ cell transplantation in males and females of fish model. Majhi and colleagues provided the first evidence of donor germ cells colonization and differentiation in the ovaries of female recipients as well as the functional viability of eggs produced by this method. Nevertheless, there was a significant difference between the sexes in the number of individuals which were successfully colonized by donor cells. For instance, 80% of male recipients had donor germ cells occupying the basal compartment of spermatogenic lobules at 8 weeks whereas only 20% of the females had donor cells in the corresponding location (the ovigerous lamellae) at the same time. It remains to be determined if such difference in implantation rates between males and females reflects histological or physiological differences between the sexes. More likely, they may be the result of subtle differences in the degree of sterility or in the microenvironment conditions of the gonads or even be an artifact due to the poor control over the number of germ cell stem cells injected into each individual. In any case, the fact that even under suboptimal conditions the donor germ cells have successfully colonized and undergone meiotic differentiation to produce gametes of donor origin in both sexes makes a powerful case for the simplicity and usefulness of intra-papillar germ cell transplantation in adult fish.

Seven months after the procedure, donor-derived gametes were detected in 17% and 5% of the surrogate male and female, respectively, suggesting that the somatic cells of recipients' gonads have supported the proliferation and differentiation of donor germ cells. Artificial fertilization using the gametes from surrogate parents resulted in normal development of embryos with no noticeable abnormalities. In this case, donor-derived germline transmission rates of 12.6%-39.7% was recorded in three surrogate male and a surprising 52.2% in the surrogate female 7 months after germ cell transplantation. Of great interest is that the rates remained relatively stable between 7 and 11 months after germ cell transplantation. These results compare favorably with germline transmission rates previously recorded for surgical germ cell transplantation in adults of our fish model (1.2-13.3%) and for transplantation of primordial germ cells into recipient embryos of salmonids (2-4%) and trout (40-46%). However, they are still well below the 100% donor-derived gametes produced by transplantation of primordial germ cells into embryos neutered by either triploidy or the use of an antisense dead end morpholino oligonucleotide. As discussed above, it is tempting to conclude that the increased germline transmission rates in this study compared to the previous studies are the result of enhanced depletion of endogenous germ cells prior to germ cell transplantation. If this can be verified, it may be possible to enhance even further the germline transmission rates by additional optimization of the recipient preparation process and perhaps reach the 100% rates obtained for embryos. More recently, Majhi's group from India had reported to obtain 100% sterility in adult common carp by injecting 5 doses of 40 mg/kg Busulfan, at 2-week intervals, during the treatment period of 10 weeks.

In conclusion, the recipients produced in the present study proved to be able to host exogenous spermatogonia and oogonia cells that were implanted non-surgically through the genital papilla and support their differentiation into viable and functional gametes. Thus, the present approach using thermo-chemical treatments to prepare adult recipients for surrogate broodstock in a short time could be a valuable alternative to methods that take considerably longer time and labor. The proposed combination of methods to prepare recipients and to perform germ cell transplantation could be valuable in cases that require immediate attention such as species facing imminent extinction and for which suitable hosts cannot be prepared in a timely fashion.

IMPORTANCE OF SURROGATE BROODSTOCK FOR AQUACULTURE

Production of genetically superior seed for aquaculture

Broodstock that are difficult to propagate in captivity or for those the breeding protocol is yet to be developed are often collected from the wild and maintained in a large land-based tank, and fertilized eggs are collected using an egg collection net set at the water outlet of the tank. However, this method is not appropriate for efficient breeding methods that often require the mating of particular parent individuals carrying desirable genetic traits i.e. selective breeding. In some fish species, males and females with superior traits can be artificially bred using a maturation induction technique using exogenous hormone administration. However, some fish are quite sensitive to handling stress, and there is also a risk of losing valuable parent fish during egg or sperm collection. Furthermore, the survival rate is usually low for fertilized eggs obtained by *in vitro* fertilization compared with naturally spawned eggs. Therefore, for fish species in which *in vitro* fertilization is difficult assisted reproduction technique can be used to produce the progeny. In group-spawning marine fishes, it is not easy to obtain fertilized eggs from a small number of selected individual fish. In many cases, mating between males and females with superior traits may become difficult because they do not reach maturity at the same time. To overcome these drawbacks, gonadal stem cells from an individual fish carrying superior traits can be transplanted into a large number of recipient fish to generate a large number of females and males that produce gametes carrying genes associated with the superior traits. Group spawning of the resulting surrogate parent fish thereby leads to fertilization involving eggs and sperm derived from donors with superior traits. As a result, the efficiency of breeding will be much improved. This concept was successfully demonstrated by the scientist from Japan by using the chub mackerel. Further, the intra-species transplantation technology can also give production of genetically superior offspring from a donor. This is particularly important for the large-bodied fish that are difficult to breed in captivity. Nevertheless, inter-species transplantation has many more advantages than intra- species transplantation. According to the Japanese research group led by Dr. Yoshizaki Goro, this technology is expected to be able to produce bluefin tuna from a small mackerel species such as the chub mackerel. Adult bluefin tuna are quite large and require 3–5 years to reach sexual maturity. In contrast, adult chub mackerel, which also belongs to the Scombridae, weigh 300 g and reach maturity in 1 year. Therefore, if the chub mackerel is able to produce eggs and sperm of bluefin tuna, the space, labour, and cost required for maintenance of the broodstock would be minimized. In addition, transplanting germline stem cells from bluefin tuna with superior traits into a small fish of the Scombridae would create genetically improved seedlings of bluefin tuna, which would be a far more efficient method than using bluefin tuna itself as a surrogate broodstock.

Production of donor-derived progeny in short duration

Many aquaculture fish species have longer reproductive maturation time. For example, Indian major carp generally attain sexual maturity at 2+ age. Thus, it aid to the investment such as food and healthcare. The aquaculture industry across the world have been demanding to reduce the generation time for breeding species. In this context, using the surrogacy technology, the production of progeny from commercially important fish species can be considerably reduced to few months instead of years. We have successfully produced eggs and sperm of goldfish by transplanting germline stem cells of this species into common carp. Similarly, Japanese group have generated tiger puffer progeny from grass puffer for aquaculture purpose. Generally, male tiger puffer require 2 years to mature, whereas female tiger puffers require 3 years to mature. Whereas, both sexes of the grass puffer can mature within a year when the water temperature and photoperiod are controlled. Similarly, it is now possible to produce eggs and sperm of Chinook salmon *Oncorhynchus tshawytscha* (which normally require 3–5 years to mature) in 1 and 2 years, respectively, when using rainbow trout as surrogate broodstock. In surrogate broodstock development, most important is selection of recipient fish. Generally, closely related fish species that is prolific breeder and for which the breeding protocol is well established should be chosen. In summary, surrogate broodstock technology can be useful to accelerate breeding and preserve the genetic information of elite germplasm. It can also be used to prevent the unauthorized production of pirated fish seedlings. Surrogate broodstock technology does not artificially manipulate cellular contents (unlike gene recombination or nuclear transfer technologies), which could make it advantageous for application on an industrial scale. However, there are few concern that has to be addressed before surrogate broodstock technology can be applied to the aquaculture industry. For example, there are few fish species those possess low concentrations of stem cells in the gonad. In such species, the use of maturing young fish or immature adult fish will allow the easy and efficient transplantation of germ cells because the concentration of stem cells is relatively high in the gonads of these fish. However, if younger fish are not available and one must use mature adults carrying large numbers of developed germ cells, germline stem cells must be enriched prior to transplantation. Several methods for isolating cell populations containing high concentrations of transplantable stem cells from mature testis have recently been developed. Further, monoclonal antibodies that specifically recognize spermatogonia containing a high concentration of germline stem cells have been recently generated. The use of such antibodies concentrates transplantable cells via magnetic activated cell sorting without the need for any expensive equipment, including a cell sorter. Further, to maximize the efficiency of germ cell transplantation, it is important to know the time of year when germline stem cells accumulate more in numbers in the gonads of donor fish so that more of stem cell can be used for transplantation.

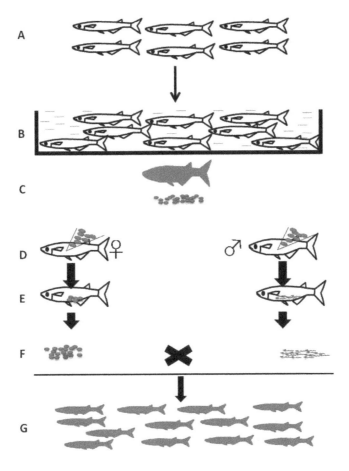

Fig. 1. Graphical illustration of surrogate broodstock development for small-bodied fishes having low gamete count. A) Selection of closely related adult recipient fish having comparatively high fecundity. B) Depletion of endogenous germ cell in recipient fish by heat-chemical treatments. C) Gonadal stem cell isolation from the donor fish (small-bodied fish having low fecundity). D) Labeling of donor cell with fluorescent dye and transplantation into the recipient gonad through genital opening. E) Months after the procedure, transplanted cells differentiated into functional gametes; F) Gametes from surrogate parent are artificially inseminated or allowed natural spawning. G) Large-scale production of donor-origin progeny.

Production of more progeny from small-bodied fish

There are several fish species, especially ornamental fishes, those are small in size and having very low fecundity. Such fishes posses very high commercial value and for their trade natural resources are continuously explored. Through surrogate broodstock technology, such fishes can be captive propagated and large number of progeny can be generated. This concept was recently successfully demonstrated by Majhi and colleagues using donor goldfish and recipient common carp model (Fig. 1).

CHAPTER-6

■——————■

INSTRUMENT USED FOR SURROGATE BROODSTOCK DEVELOPMENT IN FISH

The instrument (Fig. 1) was invented by Dr. Majhi from National Bureau of Fish Genetic Resources, Lucknow, India. The invention relates to the reproductive biotechnology of fishes. More precisely, the invention provides an apparatus for germ cell transplantation in fishes for surrogate broodstock development and general surgery purpose. The unique advantages of the instrument is that, germ cell derived from one fish species could be easily transplanted into the gonads of other fish by surgical and/or non-surgical method as the physical movement of the fish could be arrested for over 45 minutes. This is particularly important because the fish will encounter less stress and the person performing the transplantation will have enough time to transplant the cell; oppose to the other method of practice where fish is taken out of its environment and hold in hands or over a

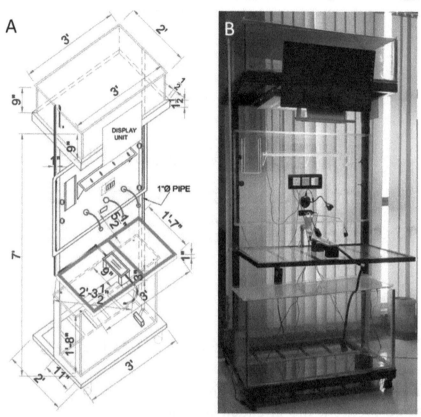

Fig. 1. View of cell transplantation instrument. The instrument shown in B was fabricated using sketch shown in A (measurement are in inch ['] and foot ["]). It consisted of two 50-L glass tanks, placed below and above a metallic frame. The centre of the frame consisted of a glass working platform. A chamber made of wood and foam was placed on the platform. The chamber was fitted with adjustable metallic clamps on both sides to hold the fish in a ventral-side-up position. The chamber has an inlet and outlet for water flow. Water containing the anesthetic agent from the lower tank was raised to the upper tank by using a water pump (0.5 hp); the water descended under gravity, passed through the operculum of the fish that was placed inside the chamber, and flowed back into the lower tank. The apparatus also has racks to hold the electrical switches, camera, and focusing light. The camera was fitted to relay the cell transplantation procedure through a display unit fitted at the top of the metallic frame. The equipment has four metallic wheels at the base to enable easy movement.

table to transplant the cell. The obvious disadvantage of doing in later way is that, while fish is taken out of water, the protective mucos layer over the skin will dry; thereby very likely the fish will get pathological infection. Further, the fish may not live beyond couple of minutes due to lack of dissolved oxygen. One more advantage of the present invention is that, the position of the fish could be kept intact without body movement. Thus, transplantation procedure will be smooth and chances of gonad rupture could be averted.

The invention also relate to the general operation practice done in fishes. The operation in fish could be for various purposes, such as placing tag and/or any other inert material in the coelomic cavity. Additionally, the catfishes those do not release spermatozoa cell in captivity could be operated to partially harvest the spermatozoa cell by surgical intervention. All these could be smoothly done using the instrument, without killing the fish. As of now, for any such operation, especially in some of the catfishes group, the male fishes were sacrificed to isolate the gonads. The instrument could also be used for general operation purposes such as display the vital organs of the fish to school or college students, without killing the animal.

By inventing the instrument for cell transplantation and surgery in fish, stem cell derived from embryos or adult fish from one fish species could be easily transplanted in to other fish gonads by surgical or non-surgical method for surrogate broodstock development that could help in conservation of valuable germlines and propagation of commercially important big size fishes those are difficult to breed in confinement due to stress of captivity.

Salient features

1. The instrument can be useful for performing cell transplantation and operation in fish, without sacrificing the fish.

2. The instrument can keep the fish under sub-conscious condition for a time period of 40 minutes, without damaging the protective mucos layer of the fish.

3. Fish of various body weight: 50-500gm and length 12-37 cm could be used for cell transplantation and/or operation in the instrument.

4. The instrument can be used for gonadal germ cell transplantation through surgical by midline incision in abdomen or non-surgical through genital opening in adult fish, without killing the animal.

5. The instrument can be used for stem cell therapy purpose to revitalize the reproductive competence of senile fish.

6. The instrument can be used for production of transgenic fish by transplantation of transfected stem cell to the host gonads.

7. The instrument can be used for investigation of internal organs of the fish by doing surgery but without sacrificing the fish.

Benefits

1. Germ cell could be transplanted into the gonads of the fish, with surgical or non-surgical method with accuracy, as the fish could remains in a fixed position under sub-conscious condition for up to 45 minutes.

2. The stress level, damage to the external and internal organs of the fish could be greatly reduced during the procedure of cell transplantation or surgery.

REFERENCES

1. Van den Aardweg GJ, de Ruiter Bootsman AL, Kramer MF, Davids JA (1983) Growth and differentiation of spermatogenetic colonies in the mouse testis after irradiation with fission neutrons. Radiat Research 94: 447-463

2. Yong GP, Goldstein M, Phillips DM, Sundaram K, Gunsalus GL, Bardin CW (1988) Sertoli cell-only syndrome produced by cold ischemia. Endocrinology 122: 1047-1082

3. Rockett JC, Mapp FL, Garges JB, Luft JC, Mori C, Dix DJ (2001) Effects of hyperthermia on spermatogenesis, apoptosis, gene expression, and fertility in adult male mice. Biol Reprod 65: 229-239

4. Billard R (1982) Attempts to inhibit testicular growth in Rainbow trout with antiandrogens (Cyproterone, Cyproterone Acetate, Oxymethalone) and Busulfan given during the period of spermatogenesis. General and Comparative Endocrinology 48: 33-38

5. Strüssmann CA., Saito T, Takashima F (1998) Heat induced germ cell deficiency in the teleosts *Odontesthes bonariensis* and *Patagonina hatcheri*. Comp Biochem Physiol 119: 637-644

6. Ito LS, Yamashita M, Strüssmann CA (2003) Histological process and dynamics of germ cell degeneration in Pejerrey *Odontesthes bonariensis* larvae and juveniles during exposure to warm water. J Exp Zool 297: 169-179

7. Ito LS, Yamashita M, Takahashi C, Strüssmann CA (2003) Gonadal degeneration in sub-adult male pejerrey (*Odontesthes bonariensis*) during exposure to warm water. Fish Physiol Biochem 28: 421-422

8. Byerly MT, Fat-Halla SI, Betsill RK, PatinÞo R (2005) Evaluation of short-term exposure to high temperature as a tool to suppress the reproductive development of channel catfish for aquaculture. North Am J Aquac 67: 331-339

9. Okutsu T, Shikina S, Kanno M, Takeuchi Y, Yoshizaki G (2007) Production of trout offspring from triploid salmon parents. Science 317: 1517

10. Arai K (2001) Genetic improvement of aquaculture finfish species by chromosome manipulation techniques in Japan. Aquaculture 197: 205-228

11. Kapuscinski AR, Patronski TJ (2005) Genetic methods for biological control of non-native fish in the Gila River Basin. Contract report to the U.S. Fish and Wildlife Service. University of Minnesota, Institute for Social, Economic and Ecological Sustainability, St. Paul, Minnesota. Minnesota Sea Grant Publication (F 20)

12. Maclean N, Rahman MA, Sohm F, Hwang G, Iyengar A, Ayad H, Smith A, Farahmand H (2002) Transgenic tilapia and the tilapia genome. Gene 295: 265-277

13. Ogawa T, Arechaga JM, Avarbock MR, Brinster RL (1997) Transplantation of testis germinal cells in to mouse seminiferous tubules. Int J Dev Biol 41:111-122

14. Scott GG, McArthur NH, Tarpley R, Jacobs V, Sis RF (1981) Histopathological survey of ovaries of fish from petroleum production and control sites in the Gulf of Mexico. J Fish Biol 18:261-269

15. Yoshizaki G, Tago Y, Takeuchi Y, Sawatari E, Kobayashi T, Takeuchi T (2005) Green fluroscent protein labeling of primordial germ cells using a non-transgenic method and its application for germ cell transplantation in salmonidae. Biol Reprod 73: 88-93

16. Blazer VS (2002) Histopathological assessment of gonadal tissue in wild fishes. Fish Physiol Biochem 26: 85-101

17. Russell LD, Griswold MD (2000) Spermatogonial transplantation-an update for the millennium. Mol Cell Endocrinol 161: 117-120

18. Swanson WF (2006) Application of assisted reproduction for population management in felids: The potential and reality for conservation of small cats. Theriogenology 66: 49–58

19. Pukazhenthi BS, Comizzoli P, Travis A, Wildt DE (2006) Applications of emerging technologies to the study and conservation of threatened and endangered species. Reproduction Fertility and Development 18: 77–90

20. Wenzhi M, Lei A, Zhonghong W, Xiaoying W, Min G, *et al.* (2011) Efficient and safe recipient preparation for transplantation of mouse spermatogonial stem cells: pretreating testes with heat shock. Biology of Reproduction 85: 670-677

21. Lacerda SMSN, Batlouni SR, Costa GMJ, Segatelli TM, Quirino BR, *et al.* (2010) A new and fast technique to generate offspring after germ cells transplantation in adult fish: the Nile tilapia (*Oreochromis niloticus*) model. PLoS ONE 5: e10740

22. Wong T, Saito T, Crodian J, Collodi P (2011) Zebrafish germline chimeras produced by transplantation of ovarian germ cells into sterile host larvae. Biology of Reproduction 84: 1190–1197

23. Brinster RL, Averbock MR (1994) Germline transmission of donor haplotype following spermatogonial transplantation. Proceedings of National Academy of Science USA 91: 11303-11307

24. Honaramooz A, Behboodi E, Hausler CL, Blash S, Ayres S, *et al.* (2005) Depletion of endogenous germ cells in male pigs and goats in preparation for germ cell transplantation. Journal of Andrology 26: 698-705

25. Majhi SK, Hattori RS, Yokota M, Watanabe S, Strüssmann CA (2009) Germ cell transplantation using sexually competent fish: an approach for rapid propagation of endangered and valuable germlines. PLoS ONE 4: e6132

26. Yazawa R, Takeuchi Y, Higuchi K, Yatabe T, Kabeya N, *et al.* (2010) Chub mackerel gonads support colonization, survival, and proliferation of intraperitoneally transplanted xenogenic germ cells. Biology of Reproduction 82: 896–904

27. Majhi SK, Hattori RS, Rahman SM, Suzuki T, Strüssmann CA (2008) Experimentally-induced depletion of germ cells in sub-adult Patagonian pejerrey (*Odontesthes hatcheri*). Theriogenology 71:1162–1172.

28. Brinster RL (2002) Germ cell transplantation. Science 296: 2174-2176

29. Trefil P, Micakova A, Mucksova J, Hejnar J, Poplstein M, *et al.* (2006) Restoration of spermatogenesis and male fertility by transplantation of dispersed testicular cells in the chicken. Biology of Reproduction 75: 575-581

30. Brinster RL, Zimmermann JW (1994) Spermatogenesis following male germ-cell transplantation. Proceedings of National Academy of Science USA 91: 11298–11302.

31. Takeuchi Y, Yoshizaki G, Takeuchi T (2003) Generation of live fry from intraperitoneally transplanted primordial germ cells in rainbow trout. Biology of Reproduction 69: 1142–1149

32. Morita T, Kumakura N, Morishima K, Mitsuboshi T, Ishida M, *et al.* (2012) Production of donor-derived offspring by allogeneic transplantation of spermatogonia in the yellowtail (*Seriola quinqueradiata*). Biology of Reproduction 86: 1–11

33. Lee S, Iwasaki Y, Shikina S, Yoshizaki G (2013) Generation of functional eggs and sperm from cryopreserved whole

testes. Proceedings of National Academy of Science USA 110: 1640–1645

34. Saito T, Goto-Kazeto R, Arai K, Yamaha E (2008) Xenogenesis in teleost fish through generation of germ-line chimeras by single primordial germ cell transplantation. Biology of Reproduction 78: 159–166

35. Takeuchi Y, Yoshizaki G, Takeuchi T (2004) Surrogate broodstock produces salmonids. Nature 430: 629

36. Nagano M, Avarbock MR, Brinster RL (1999) Pattern and kinetics of mouse donor spermatogonial stem cell colonization in recipients testes. Biol Reprod 60: 1429-1436

37. Ogawa T (2001) Spermatogonial transplantation: the principle and possible application. J Mol Med 79: 368–374

38. Ogawa T, Dobrinski R, Brinster RL (1999) Recipient preparation is critical for spermatogonial transplantation in the rat. Tissue Cell 31: 461–472

39. Lacerda SMSN, Batlouni SR, Silva SBG, Homem CSP, França LR (2006) Germ cell transplantation in fish: the nile tilapia model. Anim Reprod 2: 146–159

40. Nóbrega RH, Batlouni SR, França LR (2009) An overview of functional and stereological evaluation of spermatogenesis and germ cell transplantation in fish. Fish Physiol Biochem 35: 197–206

41. Grier HJ (1976) Sperm development in the teleost *Oryzias latipes*. Cell Tissue Res 168: 419–432

42. Wang CQF, Cheng CY (2007) A seamless trespass: germ cell migration across the seminiferous epithelium during spermatogenesis. J Cell Biol 178: 549–556

43. Somoza GM, Miranda LA, Berasain GE, Colautti D, Remes Lenicov M, *et al.* (2008) Historical aspects, current status and prospects of pejerrey aquaculture in South America. Aquac Res 39: 784–793

44. Majhi SK, Maurya PK, Kumar S, Mohindra V, Lal KK (2019). Depletion of endogenous germ cells in Striped catfish *Pangasianodon hypophthalmus* (Sauvage, 1878) by heat-chemical treatments. Reproduction in Domestic Animals 54 (12): 1560-1566

45. Majhi SK, Rasal AR, Kushwaha B, Raizada S (2017) Heat and chemical treatments in adult *Cyprinus carpio* (Pisces cypriniformes) rapidly produce sterile gonads. Animal Reproduction Science, 183: 77-85

46. Majhi SK, Hattori RS, Rahman SM, Strüssmann CA (2014) Surrogate production of eggs and sperm by intrapapillary transplantation of germ cells in cytoablated adult fish. PLoS ONE 9(4): e95294

ANNEXURE-I

PROCESSING STEPS OF FISH GONADS FOR HISTOLOGY

1. Collection of tissue: Size of the tissue less than 1cm^3
2. Fixed in Bouins' solution for 48hrs
3. Dehydration

50% Methanol	1hr
70% Methanol	1.5hr
90% Methanol	1.5hr
95% Methanol	1.5hr
Absolute methanol	1.5hr
Ethanol I	2hr
Ethanol II	overnight
Chloroform	1hr
Chloroform	2hr
Paraffin	2hr
Paraffin	2hr

4. Blocking: Working temperature 62°C. Later put the block in -20°C for further process.

5. Trimming (20 micron) and sectioning (4 micron): Keep the slides in hot air oven at 50°C for 1 hr.

Staining of slides for Histology

Xylene I	5min
Xylene II	5min
Xylene III	5min
Absolute alcohol	5min
90% alcohol	5min
70% alcohol	5min
50% alcohol	5min
Distilled water	5min
Stain in haematoxylin	10min
Wash in running tap water	2min
0.5% acid (HCl) alcohol	5 seconds
Rinse with distilled water	
Scott's tap water substitute	5min
Rinse with tap water	
Stain in 1% alcoholic eosine	5min
Absolute alcohol	5min
Xylene I	10min
Xylene II	10min

Mount the slides with DPX, allow to dry in room temperature and observe under microscope

DNA ISOLATION FROM FISH TISSUE

Reagents used for DNA isolation

1. **EDTA stock solution (250mM, pH 8.0)**

    Dissolve 4.6 g of EDTA $2H_2O$ in 25 ml of distilled water. Adjust pH to 8.0 with 1 N NaOH. Make the volume to 50 ml with distilled water and autoclave.

2. **Tris buffer stock solution (1M, pH 8.0)**

    Dissolve 12.114 g of Tris in 50 ml of distilled water. Adjust pH to 8.0 with 1 N HCl. Make the volume to 100 ml with distilled water and autoclave.

3. **Lysis buffer (TEN buffer; 50 mM Tris HCl (pH 8.0), 10 mM EDTA, 100 mM NaCl)**

    Mix 6.25 ml of 1M Tris, 5 ml EDTA (250 mM) and 0.73 g NaCl in 100 ml distilled water and make the volume up to 125 ml by addition of distilled water and autoclave.

4. **TE buffer (10 mM Tris and 1 mM EDTA, pH: 8.0)**

   Add 1.25 ml of 1M Tris to 0.5 ml EDTA (250 mM) and make the volume up to 125 ml by adding distilled water and autoclave.

5. **10% SDS**

   Dissolve 10 g of SDS in 100 ml of autoclaved distilled water.

6. **70% ethanol**

   Mix 70 ml ethanol with 30 ml of autoclaved distilled water.

7. **Chloroform: isoamyl alcohol (24:1)**

   Mix 96 ml of chloroform with 4 ml of isoamyl alcohol.

8. **Proteinase K**

   20 mg/ml in autoclaved distilled water, store at -20°C.

9. **Tris saturated phenol**

   - Take 1.25 ml of 1M Tris (pH 8.0) in a measuring flask and make the volume to 125 ml by adding distilled water.
   - Weigh 15 g of phenol and dissolve completely in a sterilized beaker containing above prepared Tris (10 mM).
   - Add hydroxyquinoline to a final concentration of 0.1%.
   - Allow the phenol and water layer to separate. Note, the lower layer is phenol.
   - After 20-30 minutes, pipette out the water layer without disturbing the phenol layer. Repeat the procedure until the pH of the phenol becomes 8.0 (checked by pH indicator paper).
   - Again add remaining 10 mM Tris, mix properly and transfer to brown bottle and store in the refrigerator until further use.

Protocol for Isolation of genomic DNA

1. **Cell lysis**

   Complete disruption and lysis of cell wall and plasma membranes of cells and organelles is an absolute requirement for genomic DNA isolation. Incomplete disruption results in poor yields.

   i.   **Sperm sample**

   - Take 200 µl of sperm sample in a separate 2 ml tube and centrifuge at 6000 rpm at 4°C for 10 minutes and decant to remove methanol.

   - Fill the tubes with normal saline and centrifuge at 6000 rpm for 10 minutes and decant the supernatant. Repeat twice or more to ensure that whole methanol is removed.

   - Suspend the cells in 500 µl of TEN buffer.

   - To each tube, add 5 µl of proteinase K and 50 µl of SDS and mix thoroughly and incubate at 55°C for 2 hrs for cell lysing.

   ii. **Fish tissue and cell culture**

   - Take 100-200 mg of tissue in a 2 ml tube and cut into small pieces using sterile forceps and scissors. Homogenize with the help of automated homogenizer or mortar and pestle.

   - Add 500 µl of TEN buffer, 50 µl of 10% SDS and 5 µl of proteinase K, mix properly by inverting the tubes several times and incubate at 55°C overnight.

   - For alcohol preserved tissues, the alcohol should be removed prior to lysis.

2. Phenol Extraction

- After lysis, add equal volume of Tris-saturated phenol to the lysate and mix well till emulsion forms.

- Centrifuge the tube at 10,000 rpm for 10 minutes and collect the aqueous phase in a fresh tube

- If the organic and aqueous phases are not well separated, then centrifuge again for longer period.

- To each tube, add equal volume of Phenol: chloroform: isoamyl alcohol (25:24:1) and mix by inverting the tube several times till emulsion forms.

- Centrifuge the tube at 10,000 rpm for 10 minutes at 4°C.

- Transfer the aqueous phase into a separate tube and add equal volume of chloroform: isoamyl alcohol (24:1), mix well and centrifuge the tube at 10,000 rpm for 10 minutes at 4°C.

- Transfer the clear aqueous phase to a separate 1.5 ml tube.

3. Isopropanol precipitation

- Adjust the salt concentration if necessary with sodium acetate (0.3 M final concentration; pH 5.2) or ammonium acetate (2- 2.5 M final concentration).

- Add 0.6 to 0.7 volumes of isopropanol to the DNA solution and mix well.

- Centrifuge the tube at 10,000 rpm for 10 minutes.

- Remove the supernatant and wash the precipitate twice with 500 µl of chilled 70% ethanol.

- Air dry the DNA pellet for 15-20 minutes.

- Resuspend the DNA in appropriate volume of TE buffer and store at 4°C.

- Quantify the DNA using spectrophotometer and check the quality by running it on 0.8% agarose gel.